TECHNOLOGY OF INDIAN FLAT BREAD
"*CHAPATTI*"

The Author

Dr, S. K. Berry has worked with CSIR- Central Food Technological Research Institute, Mysore as Senior Scientist/ Deputy Director and Head of its Regional Centre, Ludhiana, Punjab from 1983 to 2002. He coordinated and supervised the R&D and technology transfer activities of the Regional Centre and maintained liaison with State and Central Government Departments to promote development of food industry in the region. He served on the advisory and consultant assignments involving assessment of technical feasibility and financial viability of setting up food processing units and modernization of the existing units in the states of Punjab, Haryana, Himachal Pradesh and Rajasthan. Prior to this Dr Berry worked as Lecturer in Food Science & Technology at University Malaya, Kualalumpur, and University Pertanian Malaysia, Serdang (1974 to 1983), and lecturer in Chemistry in Ghana, West Africa (1966 to 1969) and D A V Postgraduate College Dehra Dun, India (1965 to 1966). Dr Berry obtained his M.Sc. (Chemistry) degree from Agra University, India (1965), M.Sc. & Ph.D. in Food Science from Leeds University, Leeds, England (1970 & 1973). His R & D pursuits comprised Chemistry of food flavours, chemistry & technology of oils and fats, fruits and vegetables; and convenience and traditional foods. The results of research work have been translated into nearly 100 publications in national and international journals and some processes for manufacturing food products. He coauthored three proceedings of Symposia/ Workshop, and a monograph on Guava in India: production, preservation & processing. He has been endowed with Fellowship of Chemical Society (UK), Institute of Food Science & Technology (UK) and Association of Food Scientists and Technologists (India).

TECHNOLOGY OF INDIAN FLAT BREAD "*CHAPATTI*"

Dr. S. K. Berry, Deputy Director (Retired)
CSIR-Central Food Technological Research Institute,
Mysore, India

SCIENTIFIC
PUBLISHERS

Published by

SCIENTIFIC PUBLISHERS (INDIA)

5 A, New Pali Road, P.O. Box 91

Jodhpur 342 001 (INDIA)

E-mail: info@scientificpub.com

Website: http://www.scientificpub.com

ISBN: 978-93-89832-45-7

eISBN: 978-93-89832-47-1

Printed in India

DEDICATION

Dedicated in the memory of my loving ***Parents*** *and Son* ***Aman***

ACKNOWLEDGEMENTS

The cooperation received from the followings is gratefully acknowledged:

1. Punjab Agricultural University, Ludhiana; ICAR-Central Institute for Post-harvest Engineering and Technology, Ludhiana; ICAR-Indian Agricultural Research Institute, New Delhi; ICAR-Indian Institute of Wheat and Barley Research, Karnal for permission to use library facilities.
2. www.google.com for literature search
3. Authors of research / review articles for providing reprints
4. Food processing machinery manufacturers for brochures on chapatti making machinery and related gadgets.

PREFACE

Wheat has sustained mankind in meeting the need for protein and energy since time immemorial. Wheat flour, whole or refined has been used variously to prepare a variety of baked products both leavened and unleavened.Chapatti is one of the unleavened products. It is flat, circular, single layered product made by rolling a ball of stiff dough of whole-wheat flour with or without the addition of salt and fat, and baked on both sides on a hot iron griddle popularly called *tawa*. It has been categorized among the flat breads both leavened and unleavened (Faridi, 1988), and has been the staple food of the millions world over, especially the Indian subcontinent for millennia. So far, the preparation of chapatti and chapatti-related products has been manual and confined to households, *dhaabas* (roadside eateries), hotels, restaurants, other catering establishments, institutions providing meals, etc. Recently, like bread, chapatti-making has been mechanized and a wide range of chapatti-making machines of different capacities have been developed and have become popular. Wheat has different species and varieties. Different wheat varieties are cultivated, and new varieties are continuously being developed for selection. Austin and Ram (1971), in an Indian Council of Agricultural Research Institute Technical Bulletin, had put together the results of their research conducted on different wheat varieties for their suitability to make bread, chapatti, biscuits, pasta products, etc. Since then a lot of research work on the development of new varieties of wheat and their quality traits amenable to make chapatti of desirable quality has been done and is continuing. The present book is an attempt to consolidate the scientific information published in Journals in Food Science and Technology, Agriculture, Biochemistry and Biotechnology, and Nutrition; technical bulletins, proceedings, reports and commercial literature on the new developments and findings related to new wheat varieties, chapatti and chapatti-related products. Chapter 1 introduces to a brief history on the cultivation and use of wheat for making of bread and flat breads like chapatti, and projects the research needs to improve the quality of wheat and chapatti. Chapter 2 discusses the physicochemical, rheological and chapatti making characteristics of different varieties of *aestivum* (bread) wheat and whole wheat flour. Chapter 3 describes developments in composite flours for chapatti. Chapter 4 describes the traditional method, test baking methods and mechanization of chapatti preparation / making. Chapter 5 covers the composition and nutritive value of whole wheat flour and chapatti, and factors affecting the nutritive value of chapatti with methods to improve the nutritive value. Chapter 6 discusses the quality aspects of chapatti, and factors influencing the chapatti quality. Chapter 7 deals with the staling aspects and extension of shelf life of chapatti. Chapter 8 presents highlights on the chapatti-related products. Chapter 9 provides the consolidated alphabetically arranged literature consulted.

Dr. S. K. Berry

M.Sc. (Chem.), M.Sc. & Ph.D. (Food Sci.),

FCS & FIFST (UK), FAFST (India).

Email: skberry59@rediffmail.com.

PREFACE

CONTENTS

Dedication *(iii)*

Acknowledgement *(iv)*

Preface *(v)*

1. Introduction **1–8**

Brief History of Wheat and Chapatti, Chapatti Nomenclature, Conversion of Wheat into Chapatti, Geographical location of Chapatti Consumers, Chapatti Variants, Chapatti Market, Chapatti Shelf-life, Research Needs

2. Wheat and Chapatti Flour **9–46**

Introduction, Development and Characterization of Indian *Aestivum,* Wheat for Chapatti Making, Wheat Quality Criteria for Chapatti, Chapatti Flour or Whole-wheat Flour, Milling of Whole Wheat Flour (*Atta*), Physicochemical and Functional Properties of Whole Wheat Flour (*Atta*)

3. Composite Flours for Chapatti **47–76**

Introduction, Tuber based Composite Flours, Other Cereal-based Composite Flours, Dietary Fibre Supplemented Flours, Legume/Pulse based Composite Fours, Oilseed Based Composite Flours, Fruits and Their Products based Composite Flours, Vegetables and Their Products based Composite Flours 74

4. Chapatti Preparation and Mechanization **77–92**

Ingredients, Traditional Method for Chapatti Preparation, Standardization of Chapatti- making Method, Mechanization of Chapatti Making, Preparation of Dough, Formation of Chapatti Dough Discs (*Raw Chapatti*), Pressing or stamping operations, Sheeting Between Set of Rollers, Sheeting by Pneumatic Pressure Extrusion, Sheeting by Mechanical Extrusion, Developments in Chapatti Making Machines

5. Nutritive Value of Chapatti **93–124**

Protein, Carbohydrates, Lipids, Minerals, Vitamins, Polyphenolics and Related Components, Volatile Flavouring Compounds in Baked *Chapatti*, Factors Affecting Nutritional Quality of *Chapatti*, Effect of Processing, Effect of Antinutrients Occurring in Wheat, Improving Nutritional Quality of *Chapatti*, Fortification with Minerals

6. Chapatti Quality **125–148**

Introduction, Physicochemical Quality Characteristics of *Chapatti*, Sensory Quality, Determination of *Chapatti* Quality, Factors Influencing *Chapatti* Quality

7. Shelf-life and Staling of Chapatti **149–172**

Introduction, Staling of Baked Products, Inhibition of Staling and Extension of Shelf-life of Chapatti, Chapatti Storage at Household Level, Chapatti Storage at Commercial Level, Use of Preservatives/Additives, Irradiation Preservation, Thermal Processing, Retort Processing, Freezing Preservation

8. Chapatti - Related Products **173–196**

***Tandoori Roti*:** Traditional Method of Preparation. Standardization of Method of Preparation. Factors affecting the *Tandoori Roti* Quality, Scope for Mechanization of *Tandoori Roti*-making

North Indian *Paratha*: Nutritive Value of Paratha, Improving the Rheological Properties and Nutritive Value of North Indian *Paratha*, Preservation, Packaging and Storage of *Paratha*

South Indian *Parotta*: Introduction, Traditional Method of preparation, Desirable Parotta Quality Characteristics, Test Baking Method, Factors Influencing the Quality of *parotta*, Production of Specialty *Parotta*, torage life of *Parotta*

Naan: Introduction and Ingredients for *naan* making, Standardization of Method for *Naan* Making, Nutritive Value of *Naan*, Reports and Patents on *naan*

Poori: Traditional Method of Preparation of *Poori*, Standardization of Process of Making *Poori*, Mechanization of *Poori* Making, *Poori* Quality Improvement, Nutrient composition of *Poori*, Shelf-life of *Poori*

Roomali *Roti*: Introduction, Traditional Method of Preparation

9. References **197–230**

Chapter 1

Introduction

Brief History of Wheat and Chapatti

The development of agriculture about 10,000 BC rapidly changed the dependence on animal food to plant foods. Wheat has been the basic staple food of majority of the civilizations of Europe, West Asia and North African regions since time immemorial. The first cultivation of wheat happened about 10,000 years ago as part of the 'Neolithic Revolution 'moving away from hunting and gathering of food to cultivation of crops (Shewry, 2009). Wheat is believed to have originated in the Vavilo Centre between the Caspian Sea, the Black Sea, and the Mediterranean. The wild ancestors of wheat traced back to the Neolithic period (7500-6500 BC) have been found in South West Asia and the Middle East. The carbonized wheat grains have been unearthed from the 3000-year-old ruins of Mohen-jo-daro in the IndusValley. The ancient civilizations of Babylon, Crete, Egypt, Greece, and Rome used wheat as one of the principal food plants, and the Chinese too grew this cereal as long back as 2700 BC. Kulshrestha (1985) has reviewed the ancient history of wheat cultivation and use in the Indian subcontinent based on both recorded and exacavated evidences.The Harappans cultivated bread wheat and other cereals at Harappa (in Montgomery district in Pakistan) and Mohen-jo-daro, Chanhu-daro, Kalibangan (in Rajasthan, India) in the Indus Valley during the Harappan Chalcolithic period (2200-1600 BC) (Randhawa, 1980). The journey of flat bread making began during the Neolithic period. For making palatable flat bread, the cereal gruel was made into flat cakes, which were baked on hot stones or directly in a fireplace (Qarooni et al, 1992). In the old kingdom of Egypt as long ago as 2500 BC, hot ashes or heated stone slabs were used to bake flat bread. Ancient Egyptian and Greeks used to eat flat bread. It is reported that later Harappans ground the wheat into flour made into dough and converted into flat breads like chapattis by baking on hot pans of clay in Rajasthan during the Harappan Chalcolithic Period. Flat metal and clay plates resembling modern *tawa* (a concave shaped iron griddle); have been found in plenty at Harappan sites suggesting that the baked chapattis may have been known during the period (McIntosh, 2008). Tandoor originated in Persia (Iran) and was brought to India via Afghanistan by Arabs way back in 3000 BC. Small mud-plastered ovens have been found in Harappa and Mohen-jo-Daro. Chapatti is mentioned in the 16th Century document Ain-i-Akbari by Abu'l-Fazl ibn Mubark, vizier of Emperor Akbar. Chapatti is now globally recognized wheat-based food product consumed by millions (Achaya, 1994).

Chapatti Nomenclature

The name *chapatti* appears to be derived from the word '*chapat*' meaning in Hindi language 'flat' or 'slap' as it likens the traditional method of flattening a ball of dough by slapping between the palms of hands. However, the term "*chapatti*" has been variously described in different English language dictionaries. A Random House Dictionary of the English language spells it as chapati or chapatti, plural ending with – tis, or –ties. It describes chapatti as "a bread made with a stiff dough of flour and water, and rolled out like a pancake and baked on a griddle". The Oxford English Dictionary spells it as chupatty (Anglo-Indian) or chowpatty or chowpattie (in Hindi pronounced as chapati). It is described as "a small cake of unleavened bread, generally made of coarse wheaten meal, flattened with the hand, and baked on a griddle". The Webster's New Twentieth Century Dictionary of the English Language - unabridged spells it as chapati or chupatty (Anglo-Indian). Here, it is described as "unleavened bread or cake made of flour, water and salt." The term bread or cake is hardly appropriate for chapatti but for the fact that chapatti is made of wheat. Truly, chapatti is a flat, circular unleavened product made by rolling a ball of stiff dough of whole-wheat flour with or without the addition of salt and fat, and baked on both sides on a hot iron griddle popularly called *tawa*. Upon baking, it gets slightly puffed and acquires a golden brown colour with a few brown spots here and there on the surface. It is, thus, a food product of its own right and has no likeness with bread or cake, which are leavened products, generally made from refined wheat flour (*maida*). Chapatti is also commonly called *roti* or *phulka* (also spelled as *pulka*) and the three names of the same product are synonyms. However, *roti* is generally thicker and larger in size than chapatti, and *phulka* is thinner and smaller in size than chapatti. In Pashto chapatti is known as *doday*. However, the word "roti" in the slogan "*roti, kapada* and *makaan*" implies food for sustenance similar to the word "bread" used in Western culture. Thus, what bread made from refined wheat flour is to the Western world, chapatti made from whole-wheat flour is to the Indian subcontinent. Notwithstanding, the chapatti is classified as flat bread in the scientific literature on breads (Qarooni et al, 1992).

Conversion of Wheat into Chapatti

The chapattis and some of its more popular variants such as *paratha, poori, naan, etc.* are normally prepared fresh and served hot at breakfast, lunch and dinner and are eaten along with other vegetarian and non-vegetarian accompaniments in households, restaurants and other catering establishments. The preparation of chapatti, since time immemorial, has been a manual operation beginning with grinding of wheat into flour in a hand driven stone *chakki*, to kneading of dough and turning a ball of dough to a thin, flat circular shape either by slapping between the palms of hands or using rolling pin and board commonly called *belan* and *chakla*, respectively. The grinding of wheat grains in stone *chakki* has been subsequently accomplished using the draught animals or hydropower, and presently electricity. There has since been a revolution in grinding wheat into flour using steel plate or roller mills having capacities ranging from a few kilograms to several tons of wheat per hour. The preparation of dough too is mechanized using manually or electrically operated dough-kneading machines. Rolling of dough into circular chapatti has also been mechanized employing manually or electrically operated presses to composite systems involving turning dough into sheet, which is then cut into circular shapes. Baking of chapatti was earlier done on clay or metal plate heated using wood fire, which has subsequently been replaced by kerosene oil and now LPG or electricity. *Tandoori roti* is baked in a *tandoor*, an oval in-ground or over-ground vertical oven, the walls of which are plastered with clay. The electrically heated metallic *tandoors* are also now available. Chapatti, like bread is now being commercially manufactured in the country using indigenously developed automatic plants. This development is now being exploited by commercial catering establishments, armed forces, large food manufacturing units, and religious establishments to cater the rising demand for ready to eat chapattis. The Golden Temple at Amritsar serving *guru ka langar* (community eating) round the clock has put up an imported chapatti making plant, having capacity to churn out 2000 to 6000 chapattis an hour (Walia. 2003). Presently, there are three or four indigenously developed automatic chapatti making units in action here with total capacity of about 11000 chapattis per hour. The *ShirdiMahakitchen* in Maharashtra has also installed an indigenous automatic chapatti making plant to serve the large gatherings of devotees every day. There are many more locations where chapattis, prepared in mechanized systems, are served as *parsad* to devotees.

Geographical location of Chapatti Consumers

The chapatti or roti as it is fondly called has been the staple food of a large segment of population of the Indian sub-continent since many centuries. Chapattis are also eaten in Nepal, Tibet, Mongolia, China, and the Near-East countries (Achaya, 1994). The other versions of this product are also found in Central Asia, parts of Africa such as Kenya, Tanzania and Uganda. Laobing flat bread serves as a local variation in China. In India more than 60 per cent of the population lives on chapattis as staple food. Punjab, Haryana, Western Uttar Pradesh, Rajasthan, and Himachal Pradesh are the regions in India, where chapatti is the predominant item in the diet. In East Uttar Pradesh, Bihar, Madhya Pradesh, Maharashtra, Gujarat, and Jammu and Kashmir, chapatti assumes equal importance to rice. Though rice has been the predominant diet in southern parts of India, West Bengal and neighboring North Eastern States, chapatti has gained popularity in recent years in these regions as well (Rao, 1982). Even the Indian subcontinent Diaspora settled in various parts of the world since long still consume chapattis as a part of their daily diet. However, some differences do exist in the recipes and methods for the preparation of chapatti among the people of different regions of India as well as different regions of other countries where chapatti is consumed.

Chapatti Characteristics

The chapatti is usually round, circular in shape, sometimes triangular or rectangular. It has a diameter ranging from 12 to 15 cm, sometimes larger or very large and is called *roti* or *rota* in some parts of the Indian subcontinent. Thickness varies from 1.5 to 4 mm and sometimes thicker ranging up to 8 mm. Chapatti has low specific volume, and is dense in texture mostly crust with little crumb. Crust is thin with large surface area, whereas crumb is small in quantity and coarse and dense with a low specific volume; thinner the chapatti, lesser the crumb and higher the crust to crumb ratio (Faridi, 1988). The desired quality parameters in chapatti are: sufficient pliability, soft texture, slight chewiness, light creamish brown colour and baked sweetish wheaty aroma. Since crust constitutes a major part of the chapatti its overall appearance in terms of colour and surface uniformity are the major factors governing chapatti quality (Rao et al, 1986b).

Chapatti Variants

There are several culinary variants of chapatti; the important ones being *phulka*, *rumali roti*, *tandoori roti*, *missi roti*, *rotlee* of Gujarat, *dopatris* of Rajasthan, etc. (Achaya, 1994). The *rotlee* is thin, whereas *rumali roti* is thinner and much bigger in size resembling a handkerchief, *Dopatris* of Rajasthan are soft, thin *rotis* that come apart as two circles because of the style of rolling of the dough. Besides wheat flour, flours of millet (*Bajra*), sorghum (*Jowar*), Maize (*Makki*), finger millet (*ragi*) and Bengal gram are also used to make *bajre ki roti, jowar ki roti, makki ki roti, ragi ki roti* and *missi roti* (a blend of wheat flour and Bengal gram flour), respectively. Though *makki ki roti and missi roti* are relished in several parts of north India, *bajre ki roti, jowar ki roti* and *ragi ki roti* are consumed in areas where *bajra, ragi and jowar* are preferentially cultivated. The fat containing products which after rolling out, are pan-fried using little fat include *parathas*. It is also spelled as *parantha*, *parotha* or parata, which may or may not be stuffed with spiced pulverized vegetables, *paneer*, eggs or minced / comminuted meat, *jaggary*, etc. *Paratha* is usually round but it can be rectangular, square or triangular in shape. The other variants that are deep fat fried include *poori*, *luchis* of Bengal, and *kachori* of Rajasthan. The fermented products include *bhatura*, *kulcha, khameeri roti*, *naan*, etc. These may replace chapatti in the dietary regimen of consumers who eat chapatti three times a day. However, it is the chapatti, *tandoori roti*, *poori*, and *paratha*, which are more frequently eaten by the majority of the population of the Indian subcontinent. *Paratha*, *poori*, *kachori*, and *bhatura* are more commonly relished in the breakfast. Stuffed *parathas* containing appropriately spiced coarsely mashed boiled potatoes, or grated cauliflower or radish or chopped green leaves of fenugreek (*methi*) are the breakfast delight of North India. In fact this gourmet has been liked so much by the consumer that a row of shops making different types of *parathas* came up in an old Delhi lane popularly called *Paranthewali Gali* (lane)more than a century ago. It has been a famous landmark of the cuisine world in India. Many a dignitaries including Pt Jawaharlal Nehru, the first Prime Minister of independent India frequented this *gali* (lane) for their *paratha* meals. Today, from Kanyakumari to Kashmir, there are hundreds of varieties of local *parathas* being served in restaurants and *dhabas*. Restaurants like "Only *Paranthas*" in Mumbai serves as many as 88 types of this favourite, while the *Paranthewali Gali* eateries offer only 14 varieties. However, the future of these eateries is being projected as shaky owing to technological innovations in food industry that provide opportunities to offer ready to eat *parathas* in a convenient frozen pack. A study by KSA Technopack, Gurgaon on ready to eat food products in India has projected a multicrore Rupees paratha market in the country.

Chapatti Market

The culinary life style of the Indian subcontinent has been changing at a fast pace. The demand for ready-to-eat chapatti and related products has witnessed a continuous increasing trend. There may be a need for the following types of chapattis to cater the market:

(i) ordinary chapatti with a shelf life of 1-2 days;

(ii) chapatti containing little hydrogenated vegetable fat or refined edible oil and added preservative for chapatti to have shelf life of 3-5 days;

(iii) chapatti with or without preservative and packaged properly for in-package sterilization to give a shelf life of 4-5 months;

(iv) frozen chapatti either half-baked or fully baked and packaged adequately to have a long shelf life.

Frozen ordinary *paratha* and stuffed *parathas* are likewise now being marketed in India and abroad. It is projected that the chapatti manufacturing business is likely to grow the same way as bread- manufacturing business grew in the western world. The present market trends show that chapatti will be the top selling processed food product of the Twenty First century in the Indian subcontinent (Bedekar, 2001). In India, we have a wide spectrum of the potential buyers of ready to eat chapatti and related products. These include:

(i) households where both husband and wife go to work;

(ii) elderly couples who live alone and unable to make chapattis;

(iii) college students who cannot afford to eat in restaurant/dhaaba or canteen;

(iv) new job seekers who throng the cities;

(v) travellers;

(vi) government and non-government organizations which are entrusted to provide ready to eat foods in disaster management, etc., and

(vii) the armed forces on the war front (Bedekar, 2001).

On the other hand, there are other establishments that are required to cater a large number of inmates, such as hostels, hospitals, training institutions, industrial canteens, central jails, armed forces, religious institutions that serve *preeti bhoj* or *langer*, have become the potential buyers of chapatti making machines. A beginning has already been made in this direction as stated above. We have already entered the era of markets flooded with appropriately packaged ready-to-eat chapattis and related products.

Chapatti Shelf-life

The chapatti if stored beyond 1-2 days under ambient conditions, suffers from such defects as hardness / brittleness, off flavour, and mold growth. Actually, chapatti loses its softness and becomes hard with in 1 to 2 hours of baking (Austin and Ram, 1971). Retrogradation and staling are the phenomena responsible for decreased acceptability during storage. The preservation of the fresh like quality of chapatti thus has been a necessity, and has so far been effected to some extent by:

(i) manipulating the processing techniques,
(ii) use of suitable preservatives,
(iii) improved packaging techniques and storage conditions,
(iv) irradiation,
(v) thermal processing,
(vi) freezing preservation, etc.

Nevertheless, much research still needs to be done to find ways to provide fresh-like quality to the ready- to- eat chapatti, the staple food of the millions in the Indian subcontinent and the Indian Diaspora world over.

Research Needs

Though ready to eat chapatti is gaining popularity among the needy consumers, its fresh-like eating quality is yet to be achieved fully. A lot of R&D efforts have been made to improve the eating quality, and enhance the storage life of chapatti. The R&D frontiers that have caught the attention of the researchers include:

(i) development of suitable wheat varieties that are rich in positive chapatti-making quality factors;
(ii) supplementation with flours of other cereals or legumes;
(iii) addition of suitable additives;
(iv) addition or selective removal of certain wheat flour constituents;
(v) parameters related to dough kneading and chapatti baking; and
(vi) packaging and storage conditions.

The interaction of wheat flour constituents during kneading and resting of dough and the process of baking need to be further looked into thread bare to achieve the objective fully. Further research is required to investigate the role of chapatti dough volatiles and starch plus gluten in the development of typical wheaty aroma through non-enzymatic browning reaction between free amino acids and sugars in the chapatti dough during baking of chapatti as investigated for pan bread (Berry, 1973; Berry and Gramshaw, 1988). Development of the typical wheaty aroma needs to be investigated to improve the palatability of chapatti with reference to increased shelf life. Also the role of whole wheat flour bran in the formation of volatiles contributing to wheaty aroma needs to be researched upon. Differences between the wheaty aroma of chapattis made from white flour (*maida*) and whole wheat flour may guide us in this direction.The mechanism of staling or retrogradation, and its inhibition also needs to be investigated more thoroughly although some efforts have been made in this direction (Nanjappa et al, 1999; Shaikh et al, 2007 and 2008; Ghodke, 2009; Hemalatha et al, 2014).

Chapter 2

Wheat and Chapatti Flour

Introduction

Wheat continues to be one of the world's most important grains, especially as food supplying energy and nutrition. Billions of people world over consume wheat and demand is increasing in countries undergoing urbanization and industrialization (Shewry and Hey, 2015). India is the second largest wheat producer in the world; produced about 98.51 million MT in the crop year 2016-17 (Ministry of Agriculture, New Delhi, India). Wheat is the dominant crop in the Northern Plains of India comprising largely the states of Uttar Pradesh, Punjab, Haryana, Madhya Pradesh, Rajasthan and Bihar. Large number of end-use products such as *chapatti*, *bread*, *biscuits*, *cakes* and a variety of pasta products are made from wheat. Approximately 85 to 95 % of wheat produced in India is consumed in the form of chapatti- an unleavened flat hot plate baked product made essentially of whole wheat flour (Bedekar, 2001).The chapatti is principally prepared from the cereal wheat. Besides wheat, maize, sorghum, millet, and *ragi* are also used to make *roti* / chapatti in regions where these cereals are grown preferentially due to climatic conditions.

There are about fifteen recognized species of wheat, of which three main species of wheat, namely *Triticum aestivum, Triticum durum and Triticum dicoccum* are of some commercial significance and are cultivated in India and many other parts of the world. However, for the purpose of trading, wheat is classified into relatively distinct categories of grain hardness (soft, medium hard, hard), colour (red, white, amber), and growing habitats (spring or winter). About 95% of the wheat produced in the world is *Triticum aestivum* utilized mainly as flour (whole or refined) for the production of a large variety of leavened baked products and unleavened flat breads. It is the *T. aestivum* which is preferred for chapatti making, whereas *T. durum* is largely used for making semolina, pasta products and other alimentary pastes. The *T. dicoccum*, on the other hand, is commonly used in southern parts of India for the preparation of breakfast foods such as *upuma*, etc. There are quite large differences in grain composition and processing quality among wheat cultivars within a species. The *aestivum* wheats vary markedly for their chapatti-making properties. The hard amber-coloured Indian wheats of *sharbati* type in the olden days used to make relatively soft, silky and pliable, and well-puffed chapattis. On the contrary, the imported hard-red American and soft-white Australian wheats did not yield chapattis with these pleasing characteristics of appearance and texture. Similarly, when the species *T. durum* and *T.*

dicoccum wheats are used for chapatti making, dry and hard-textured products result which are perceived to be leathery upon chewing (Austin and Ram, 1971).

Development and Characterization of Indian *Aestivum* Wheat for Chapatti Making

The quality of milled products particularly flour is closely related to the quality of wheat and varies from one class of wheat to another depending on the end use. Numerous varieties / cultivars of *aestivum* and *durum* are known and new ones are continuously being developed and commercialized. As early as 1893, wheat cultivation trials at College of Agricultural Research, Lyallpur, undivided Punjab, the principal wheat growing area in India, showed that imported varieties of wheat would not grow successfully. It was, therefore, decided to isolate pure lines from local wheats in 1905 at the Imperial Agricultural Research Institute, Pusa, Bihar, India. Pure line selections were used until 1926 when hybridization was introduced (Aziz, 1960). Singh and Bailey (1940) at Minnesota, USA evaluated 31 wheat varieties grown during crop season 1935-36 at Lyallpur in Punjab for their chapatti-making quality and reported wide differences in softness, pliability, sweetness and flavour of chapattis. They attributed these differences to the quantity and quality of protein, and amount of sugars present in the wheat and that generated due to the diastatic activity commencing when flour is mixed with water into dough. They further observed that the chapattis retained soft texture during storage because of the good water holding capacity of the proteins in these wheats. As reported by Austin and Ram (1971), Chaudhury Ram Dhan Singh and co-workers (Annual Report, College of Agricultural Research, 1946) investigated the chapatti-making quality of improved wheat varieties grown in Punjab under different conditions of location and irrigation. They reported a wide variation (80 to 94 per cent) in water absorption of whole wheat flour and the quality of chapatti, in terms of colour, texture and sweetness, from one wheat variety to another and within the variety grown at different locations. After independence in 1947, the institute was redesignated as Indian Agricultural Research Institute, New Delhi. In 1965, "All India Coordinated Wheat Improvement Project" was created at the Institute to accelerate the varietal development programme. The objective was to develop wheat varieties that not only suited bread making but were good for chapatti making as well (Austin and Ram, 1971).

Earlier, Aziz and Bhatti (1960) considered the following characteristics as desirable in relation to the chapatti-making quality of wheat:

(a) 1000 kernel weight, which should be high (35 to 40 g).

(b) The grains should be plump and well-filled, yielding on milling a low quantity of bran.

(c) The grains should be amber coloured so that the whole wheat flour is of creamy white colour, and contains no brown or dark particles.

(d) The protein content should not be more than 11 per cent from the baking technological point of view but should be high for nutritional reasons

(e) The gluten should not be too strong as no significant fermentation takes place in the dough for making chapattis.

(f) Water absorption should be high and the dough should not dry up too much when rested so that sufficient moisture remains in the chapatti for it to retain flexibility and softness, and not become brittle.

The extensive chapatti making tests conducted at the Indian Agricultural Research Institute, New Delhi have subsequently shown that the wheat flour used for making chapatti should have high water-absorbing capacity and should make dough which is not sticky but is strong, pliable, elastic and extensible. The chapatti made from the dough should puff fully and rapidly while baking and should be sweetish in taste. It should have good keeping quality retaining the soft, smooth and pliable texture for at least 4 to 6 h (Uprety and Austin, 1969). Austin and Ram (1971) compiled a monograph entitled "Studies on chapatti-making quality of wheat" wherein they presented a standard method for chapatti preparation, standards for chapatti quality, and reported the chapatti-making quality of a number of indigenous and improved wheat varieties. They confirmed the earlier findings of Singh and Bailey (1940) that the chemical characteristics of wheat e.g. sedimentation value, Pelshenke value, and protein content influence the textural characteristics, while the sugar content and diastatic activity determine the taste and flavour of chapatti. Following characteristics were reported as desirable for a wheat variety to qualify as chapatti wheat:

Property		**Value**
Protein	%	10 to 13
Sugar	%	> 2.5
Diastatic activity	mg/10g flour	> 150
Sedimentation value	ml	30 to 37
Pelshenke value	min	120 to 150

Tall Indian wheats compared to short Mexican wheats were found to perform better in making chapattis of good quality (Abrol, 1971). Tall Indian wheats, e.g. variety C306, were acceptable to the consumer for the following chapatti-making characteristics:

(a) The grains were bold, lustrous and of amber colour.

(b) The whole meal flour was creamish in appearance.

(c) The dough possessed creamy colour, which was retained on keeping for a long period.

(d) These wheats yielded chapattis which were silky in appearance and feel, were sweet and palatable,and possessed good puffing characteristics.

The wheats which were introduced subsequently, e.g. *Kalyan Sona*, on the other hand, did not receive consumer acceptance for the following reasons:

(a) The grains were dull amber to reddish in colour.

(b) The whole meal of these wheats possessed chalky white appearance.

(c) The dough prepared from the meal possessed poor keeping quality because of darkening and early onset of fermentation.

(d) The chapattis from such wheats were brownish in colour, leathery in texture, and tended to become brittle faster on keeping.

Sinha et al (1968) studied the physicochemical characteristics of improved varieties of Indian wheats (NP series plus C273) and found the varieties NP 825, NP778 and NP 815 yielding organoleptically superior chapattis showing good relationship with the sedimentation value and farinographic properties. Abrol (1972) in a study comprising three dwarf wheats (Lerma Rojo, Sonalika and Sharbati Sonora) and three tall wheats (C 306, C 273 and C 591) observed that the *chapattis* made from freshly prepared dough of all the varieties were comparable in 'whiteness'. However, the chapattis prepared from dough kept for 2 to 4 hours at 32 ± 2^{0}C and chapattis kept overnight at room temperature tended to discolour in case of dwarf wheats. Uprety and Abrol (1972) pointed out that most of the varieties with acceptable palatability were having high damaged starch content. The authors suggested that damaged starch together with the sugar content should be used as a selection tool for *chapatti* quality. Tikoo et al (1973) too reported that the chapattis prepared from the dough of some of the introduced Mexican dwarf wheats tend to darken earlier on the baking griddle (*tawa*) itself than those of tall varieties. Sinha and Singh (1974) carried out studies on the physical, milling and farinographic characteristics, and gluten content and sedimentation value of 25 improved commercial wheat varieties from four different states of India. Based on results, 6 varieties were classified as bread wheats, 14 varieties comprising mostly hard to semi-hard and soft wheats fit for either chapatti making or for biscuit, etc., and 5 varieties of durum wheat suitable for making *suji* (*rawa*) and pasta products. According to them Sharbati Sonora, Sonalika, *Kalyan Sona* and C273 were suitable for bread making while most of the varieties from Madhya Pradesh were suitable for chapatti making. Joshi et al (1975) reported regional variations in the farinographic and other quality characteristics of *Kalyan Sona* variety of wheat and concluded that wheat of one and the same variety when cultivated under different agro-climatic conditions may not yield a produce of uniform quality, but may differ in quality characteristics. Chopra and Bhatt (1975) evaluated three high yielding wheat varieties viz. WG357, WL212 and *Kalyan Sona* along with *desi* variety C 306 for protein and gluten content, and organoleptic properties and keeping quality of chapattis. Though the gluten content was highest in C306, the gluten from WG357 had the highest baked volume suggesting that baking quality does not depend entirely on the protein content but the quality of gluten is also important. The differences in the acceptability of chapattis prepared from different varieties were, however, non-significant.

Shurpalekar *et al* (1976) tested the chapatti-making quality of 17 varieties of Indian *aestivum* wheats which contained protein in the range between 8.1 and 13.5 per cent with damaged starch content and diastatic activity rather on the lower side perhaps because hammer mill was used to grind the wheat. Of the 17 wheat varieties, *Sharbati Sonora, Sonalika, Kalyan Sona*, K68, D134, DGPR-LB, DGPR65, Raj821, Raj842,

Raj857,WG377, and WG357 yielded chapattis with overall acceptability good to excellent though farinograph water absorption for chapatti dough ranged between 46.1 and 57.5 per cent, which was considerably lower than reported by Austin and Ram (1971).This anomaly may perhaps be attributed to the use of hammer mill to produce whole wheat flour, which contained less of damaged starch and hence lower diastatic activity, consequently low water absorption. The variety Raj 842, however, was outstanding, possessing all the qualities required for both bread as well as chapatti making. Shurpalekar and Prabhavathi (1976) assessed six Indian wheat varieties for chapatti- making quality and showed that the varieties *Kalyan Sona* and K68, with protein contents of 8.7 and 10.7 per cent and starch damages of 6.6 and 6.8 per cent, respectively, were excellent, while the soft wheat variety *Chhoti Lerma*, with 8.8 per cent protein and 2.8 per cent starch damage, was fairly acceptable, and the varieties *Sharbati Sonora* and *Sonalika,* with protein contents of 11.1 and 11.5 per cent and starch damages of 4.5 and 4.8 per cent, respectively, gave good chapattis.

Bakhshi et al (1979) tested 36 improved strains of wheat grown at five different locations for chapatti-making qualities and found that location significantly affected protein content, water absorption, dough- handling, and puffing characteristics, taste, flavour and texture of chapattis. The correlation studies showed that the strains having better dough-handling properties puffed well and had better taste and flavour. Further, Gill *et al* (1980) evaluated some of the promising strains for *chapatti* quality and found a few of them to be better than the existing check varieties. At Haryana Agricultural University, Hisar, several high yielding wheat varieties were tested for their chapatti-making quality (Popli and Dhindsa, 1980; Dhingra *et al*, 1982; Taneja *et al*, 1983). The varieties WH147, WH157, WH283, WH291, HD2281, C306, and *Sonalika* were found suitable for chapatti making. Ram and Nigam (1981) tested five wheat varieties viz. Sonalika, HD1949, HD4530, K65 and K68 for physicochemical and chapatti characteristics in relation to varietal differences in gluten composition. The variety HD 4530 had high protein content and high water absorption, and yielded chapattis fully puffed, soft and pliable, and judged best in overall eating quality (**Table 1**).

Table 1: Physicochemical and chapatti characteristics of the five wheat varieties.

Wheat variety	Protein Content	Water absorption	Chapatti Characteristics		
			Puffing	Texture	Acceptability
HD4530	16.5	70.0	Rapid, full	Soft & pliable	Excellent
HD1949	14.8	70.2	Slow, partial	Slightly tough & leathery	Poor
K65	11.3	66.0	Slow, full,	Slightly leathery	Good
K68	12.3	68.0	Slow, full	Slightly leathery	Good
Sonalika	12.5	67.8	Rapid, partial,	Slightly stiff	Fair

Source: Ram and Nigam (1981).

Singh *et al* (1983) at Punjab Agricultural University, Ludhiana tested 18 promising strains of wheat along with four commercial varieties for their chapatti, bread and cookie-making quality and rated the strains WL2549, WL3282, WL3720, WL4151, WL4232, WL4234, WL4382, WL4478, WG2029, WG2080, WG2120, WG2142, and

WG2203 very good for chapatti making, matching the commercial varieties WG357, WL711, WL1562, and *Sonalika*. Adsule *et al* (1985) evaluated four promising cultivars of wheat from Maharashtra, *viz.* NI 8188, NI 8223, NI 8289 and NI 8303 and compared with existing varieties namely, NI 5439 and HD 2189 for chapatti-making quality. The new varieties were found superior to HD 2189 but inferior to NI 5439 for making chapattis of acceptable quality. Rao *et al* (1986b), while developing chapatti test-baking method, tested the chapatti-making quality of six commercial Indian wheat varieties and reported that the soft wheat variety Pissi-local with low water absorption capacity of 67.8 per cent possessed poor chapatti making quality, whereas other varieties (WG 357, Sharbati Sonora, Kalyan Sona, Punjab) with high water absorption capacity (73 to 76.6 per cent) resulted in good quality chapatti. Das et al (1990) reported that among the new varieties Ajantha and PBN 135 had the highest protein content, and Ajantha having the highest wet gluten content. The new varieties as well as varieties N59, Kalyan Sona, NI 5439 and Sonalika received a good score for overall chapatti acceptability. Chapattis made from NI 5439, PBN 135 and Ajantha wheat varieties were judged the best after storage.

Time and again it is said that the medium strong wheats make chapatti of good quality. Contrary to this some weak and strong wheats too make chapattis of good quality. For example, the quality characteristics of the variety C306 (**e.g. protein 10.71 per cent, sedimentation value 40 ml, wet gluten 26.9 per cent, water absorption capacity 67 per cent, stability 3.3, etc**.) fail to qualify it to be a member of medium strong category. It is the weakest material among some of the commercial medium strong varieties but still makes the best chapattis (Misra, 1998). Singh et al (1999) tested 20 advanced wheat genotypes for physicochemical and rheological characteristics. The majority of the varieties studied were characterized as medium protein, medium to high content of gluten of poor strength in their flours. Sugars content was normal but majority of the varieties were low in diastatic activity. The farinographic properties such as farinograph water absorption, dough development time, dough stability and mixing tolerance index, indicated week nature of the dough produced by flours from these genotypes. Bansal *et al* (2000) studied the suitability of advanced wheat genotypes for bread, biscuit and chapattis and reported some new genotypes to be better than the check varieties but none could surpass the best chapatti wheat, C 306.The authors observed that high water absorption of dough and high moisture content of chapattis contributes to their softness.

Prabhasankar et al (2002) attempted to relate physicochemical and biochemical characteristics of 9 wheat varieties (GW190, HD 2189, HD 2501, MACS 2496, NIAW 34, NIAW 301, NIAW 5439, PBW 173 and Sonalika) grown at one location to chapatti making quality. Amongst the physicochemical characteristics, grain hardness, hectoliter weight, ash content, protein content, damaged starch, sodium dodecyl sulphate (SDS) sedimentation value and falling number ranged between 9.1 to 11.6 kg, 72 to 84 kg, 1.4 to 2.1 per cent, 12.2 to 16.2 per cent, 14.6 to 23.3 per cent, 59 to 76 ml and 347 to 684, respectively. The chemical characteristics tested indicated that most of these varieties belong to medium strong wheat type with low α-amylase content. While among the rheological properties of these wheat varieties water absorption, dough development time, dough stability and mixing tolerance index ranged between 74 to 81 per cent, 5 to 7.5 min, 2.5 to 5.5 min and 20 to 60 BU. The variety Sonalika had the minimum water absorption and the variety NIAW 34 had the maximum. There were significant correlations between some flour quality parameters and chapatti quality, which may enable to draw flour specifications for producing chapattis of desired quality. Damaged starch and chapatti water absorption are the critical factors in determining the quality of chapatti. The studies concluded that medium strong wheat varieties are suitable for chapatti making.

Singh *et al* (2006) evaluated 16 wheat varieties grown in North- Western parts of India for their physicochemical, milling, and end use quality and reported that varieties C 306, Raj 3765, PBW 373, WH 147, WH 157 and UP 2338 were good for chapatti making .Mohan and Gupta (2011) studied a number of processing and nutritive qualities in 10 popular bread wheat varieties (WH 542, HD 2687, PBW 343, PBW 502, PBW 550, DBW 17, Raj 3765, PBW 373, DBW 16) of Indo-Gangetic plains at five different locations for differences among varieties, locations and crop seasons. Variety-location interactions were found largely insignificant but year x variety registered importance for a number of quality characters. Location effect was not noticed in, flour recovery, gluten index, iron content and chapatti quality. The suitability of the varieties DBW 17 and PBW 550 in the favourable environments existing in the North-Western Indo-Gangetic plains was projected for their potential in combining high yield with acceptable wheat quality.

Mehta *et al* (2006) studied the effect of farmyard manure and nitrogen application at different stages of crop growth on the chapatti quality. An improvement in the chapatti quality was reported through the application of farmyard manure and nitrogen at later stages of crop growth.

Navnidhi et al (2009) tested 12 wheat varieties for physicochemical properties and chapatti making quality. The chapattis prepared from these wheat varieties were evaluated for puffing height, texture, and sensory characteristics like colour, shape, nature of spots on surface, pliability, aroma and taste, mouthfeel, and handfeel. The sensory characters of chapatti were correlated with the physicochemical parameters of the wheat varieties for rating their chapatti making quality. The varieties C 306, HD 2687, GW 381, HW 2004, PBW 175 and WH 912 were rated as good chapatti varieties, whereas HD 2189, HUW 598, NI 5439, WH 291, Sonak, and UA 5410 were rated as poor chapatti varieties. Jha et al (2010) evaluated the rheological and baking characteristics of 17 winter x spring cross lines along with five existing BW lines grown during 2007-08 and 2008-09 seasons and concluded that overall BW 9134 and BW 9136 produced consistently good chapattis.

Chaudhry et al (2016) evaluated four wheat varieties viz. C 306, HI 977, DW 16 and HW 2004 for physicochemical, gluten, rheological and chapatti quality characteristics (**Table 2**). The size exclusion chromatography of these varieties showed variation in their relative proportion of polymeric glutenins, monomeric gliadins, and low molecular weight monomeric proteins. Such variations showed marked association with gliadin/glutenin ratio, resistance/extension ratio, gluten index, SDS sedimentation value, dough development time, dough stability, etc. as well as chapatti quality characteristics. The correlation studies revealed that harder wheat variety with overall high protein content and higher relative proportion of gliadins than glutenins as in variety C 306 excelled in chapatti making as gliadins appropriately control dough stretchability resulting in full puffing of chapatti with optimum chewiness and lesser tearing resistance. On the other hand, strong wheat varieties viz. HI 977 and DW 16 having higher amount of glutenins were not found suitable for chapatti making as the dough tends to roll back during sheeting, and resulted in chapatti with greater chewiness and higher tearing resistance. The soft wheat variety viz. HW 2004 with low protein content and containing exceeding levels of gliadins than glutenins resulted in excessively flowy chapatti which offered poor mouthfeel and chewability.

Table 2: Physicochemical, gluten, rheological and chapatti quality characteristics of some wheat varieties.

Characteristic		Variety			
		C 306	DW16	HI 977	HW2004
Grain hardness	N	87.27	68.43	76.30	75.07
Damaged starch	%	6.87	5.92	6.56	6.13
Falling number	sec	536	501	501	471
Protein content	%	12.56	11.59	12.43	8.68
SDS value	ml	45	46	62	35
Dry gluten	g/100g	11.7	9.6	8.9	7.8
Resistance/Extension		0.38	0.58	1.49	0.78
Gluten index		62.9	77.8	98.7	61.6
Gliadin	g/100g	6.1	5.3	4.3	6.5
Glutenin	g/100g	3.9	4.7	5.7	3.5
Gliadin / Glutenin ratio		1.58	1.14	0.7	1.83
Water absorption	%	62.40	53.73	56.73	55.97
Dough development time	min	2.98	3.27	7.38	1.89
Dough stability	min	4.14	5.84	10.80	4.06
Puffing height	cm	6.0	5.5	5.8	5.7
Compression force	g	54.96	89.90	82.03	65.10
Chapatti score		93.67	83.67	75.00	91.33

Source: Chaudhry et al (2016).

Kundu et al (2017) evaluated 50 wheat varieties for chapatti quality based on grain characteristics, dough rheological and pasting properties, and sensory characteristics. Majority of the wheat varieties were found to be medium hard to hard based on kernel texture. The damaged starch content and water absorption capacity were found to be the important parameters for chapatti quality as both parameters had significant positive effect on the pliability and puffing height of chapatti as both these chapatti characters are much sought after by the consumer. The protein content and gluten strength parameters viz. SDS sedimentation value, dough stability and gluten index were found to have negative impact on chapatti quality resulting in chapattis with hard texture which affects product overall acceptability adversely. Less than fifty per cent of the varieties studied were found to be good to acceptable for chapatti making suggesting the need for developing product-specific varieties.

Survase et al (2017) evaluated and compared 9 wheat varieties (8 Indian and 1 Australian) for chapatti and bread making quality. The extensibility of chapatti correlated positively with chapatti scores for overall quality, pliability, handfeel, chapatti eating quality, and taste. Overall chapatti quality also correlated with resistance to extension as measured on texture analyzer. The high molecular weight glutenin subunit protein composition assessed by SDS polyacrylamide gel electrophoresis did not correlate with the overall chapatti score. A negative correlation was found between chapatti and bread scores suggesting different requirements for chapatti and bread quality.Recently, Panghal et al (2019) have characterized several wheat varieties for their chapatti-making quality. The physicochemical characteristics like thousand kernel weight, hectoliter weight, SDS-sedimentation value, ash content, copper content, and iron content of wheat varieties tested were found to be significantly correlated with chapatti quality. The wheat varieties with good and poor chapatti qualities were identified based on the results. The wheat varieties C 306, GW 190, GW 322, GW 381, HD 2687, HW 2004, K 8027, K 9107, LOK 1, PBW 175, UP 1109, UP 2382, UP 262, UP 2425, and WH 912 received high chapatti quality scores, whereas wheat varieties HD 2189, HUW 598, NI 5439, NIAW 34, PBW 343, PBW 373, RKD 130, SONAK, UAS 410, WH 147, WH 157, WH 711, WH 773, and WH 291 received low quality scores being of poor quality. This study is helpful in understanding the diversity of wheat varieties with reference to their chapatti-making quality. The identification of wheat varieties receiving the highest chapatti quality score may prove beneficial to food industry using fully mechanized process to produce good quality chapatti and related products.

In the past, greater attention in wheat-breeding programmes was paid to develop varieties / cultivars that are high yielding and resistant to disease and insect attack. Currently, increasing emphasis has been placed on the quality of the grain for processing, especially the milling quality and the baking, chapatti-making and pasta products-making (for durum wheat) quality. Genetic manipulation to alter grain polyphenol oxidase activities, starch composition, gluten strength, and tolerance to postharvest sprouting, etc. is receiving priority. The production of old varieties is usually discontinued as soon as new varieties of equally satisfactory agronomic characteristics and of superior processing quality are made available (Proceedings of the 55th All India Wheat and Barley Workers meet held at CCSHAU, Hisar from 21-24 August, 2016. All India Coordinated Wheat and Barley Improvement Project).

Since 1965 till date, a total of 448 wheat varieties have been released in different regions of India, of which 378 varieties are bread wheat and rest durum, dicoccum and triticale varieties. Though all bread wheat varieties can be used to make chapattis, but the quality of chapatti differs tremendously (Refer Chapter 4 for chapatti quality characteristics). Among the bread wheats following varieties listed in **Table 3** have been characterized for their Chapatti-making quality traits from good to excellent. Their distinguishing features particularly in relation to yield, agronomic requirements and their reactions to biotic and abiotic stresses have also been documented (Gupta et al, 2018). Since then 19 more wheat varieties have been developed and released.

Table 3: List of Bread Wheat varieties notified to have chapatti making quality.

Variety	Quality	Variety	Quality	Variety	Quality
C 306	Excellent	HDR 77	Good	LOK 1	Excellent
CPAN 1676	Very good	HI 784	Good	MACS 6478	Good
DBW 168	Excellent	HI 1500	Good	MACS 2496	Good
DL 788- 2	Good	HI 617	Good	MP 3382	Excellent
DWR 39	Good	HI 1605	Excellent	NI 747-19	Very good
HB 208	Good	HI 1479	Good	PBW 65	Good
HD 2270	Good	HI 1454	Good	PBW 660	Excellent
HD 1941	Good	HI 977	Good	RAJ 1482	Good
HD 2236	Good	HP 1744	Good	RAJ 2184	Good
HD 2833	Excellent	HPW 349	Good	RAJ 4238	Good
HD 2987	Good	HS 542	Good	WG 357	Excellent
HD 2009	Good	K 1317	Excellent	WL1562	Good

Source: Gupta et al (2018).

Several authors have earlier studied the physicochemical characteristics and chapatti making quality of several bread wheat varieties other than listed above. Some of these include: WG 367, WG 377, NP 824, PB.C518, PB.C591, PB.C281, PB.C273, Raj 3077, HD 2329, HD 4530, CPAN 3004, NP 4, PBW 175, Raj 4120, K 0307, K 8027, HD 2888, CW 322, HI 1531, HW 2004, K 9107, GW 322, HI 1544, PBW 509, PBW 527, PBW 226, PB.C591, HI1418, More and more newer varieties/cultivars are being developed with improved quality traits in different climatic zones of India.

Wheat Quality Criteria for Chapatti

Wheat quality is essentially a relative concept usually judged by its suitability for a particular food product such as bread, chapatti, biscuits, cake, pasta products, etc. The wheat which is suitable for one type of food product may be quite unsuitable for the other. Among the bread type wheats, the grain quality required for chapatti making is different from that required for pan bread. Extensive studies carried out on different quality parameters have shown that there are wide differences in wheat varieties, which are attributed mainly to their genetic makeup and environmental factors. Kundu et al (2016) evaluated different wheat cultivars grown at different locations for grain quality and whole wheat flour (WWF) characteristics. The grain characters like dimension, thousand kernel weight and hardness showed significant variations among the cultivars. The damaged starch content of WWF showed a highly positive correlation with the grain hardness. The wheat variety HD2932 exhibited wide diversity in protein content as well as quality. The chapatti making potential of wheat is derived largely from the quantity and quality of its protein (Austin and Ram, 1971). Protein quantity is influenced by environmental factors, while the quality of the protein is genetically determined. Indian wheats are generally exposed to high ambient temperatures during the grain filling / ripening period which not only affect protein synthesis but also damage the elastic properties of the gluten proteins resulting in dough having higher strength and less elasticity. As a result most of the Indian wheat varieties come under medium hard category having lower content as well as inelastic gluten which makes them unsuitable for bread making but good for chapatti making (Singh, 2001). In wheat varieties that are grown under comparable environmental conditions, high quality wheat will produce good chapatti over a fairly broad range of protein levels whereas poor quality wheat will yield relatively low quality chapatti even at high protein contents.

Also, the wheat may get damaged due to many causes in field before harvest, during harvest or during artificial drying operations, subsequent storage and handling. Blight, scab or fungal attack, immaturity at harvest, sprouting in field due to wet weather, frosting when wheat is still immature, breakage during thrashing, excessive spontaneous heating during bulk storage, high moisture during storage, insect infestation, etc. all result in loss in quality of wheat, which may get reflected in the end product. Infection of wheat grains with Hill Bunt and Karnal Bunt has been reported to impair the quality of chapattis. It has been observed that 0.5 per cent infection due to Hill Bunt and 1.0 per cent due to Karnal Bunt is sufficient to render the wheat grains unpalatable for human consumption owing to fishy odour imparted to the whole wheat meal (Mehdi et al, 1973). Therefore, the wheat to be used for milling into flour should possess good appearance in terms of grain size, shape, soundness, colour and luster; free from diseased, sprouted and weathered grains; free from chaff, straw, stones, soil, weeds, ergot and other foreign matter; and fungus and insect attack. Grain appearance is important in grain trade. Further, the wheat should meet the following physical and chemical criteria of quality (Pomeranz, 1988).

Physical Quality

- ***Hectolitre weight*:** Hectolitre weight is the weight of wheat occupied in a space of 100 litres. High test weight wheats generally yield same quantity of flour than wheats of lower test weight. Weathered and infested and insect damaged grains have less test weight. Hectolitre weight is an important parameter for millers as it is positively correlated with flour recovery.

- ***Kernel weight / Thousand kernel weight*:** This is usually expressed in terms of weight per thousand kernels. It is a function of kernel size and density. Large and dense wheat kernels have higher proportion of endosperm as compared to less dense ones. Kernel weight is the index of flour yield; higher the kernel weight greater the flour yield.

  Wheat varieties with high thousand kernel weight and hectolitre weight yield good quality chapattis (Panghal et al, 2019).
- ***Kernel size and shape*:** Different varieties of wheat vary in kernel size and shape. Kernels are passed through sieves and percent overs are determined. Kernel size reflects kernel weight and is index of flour yield; greater the kernel size higher the flour yield.
- ***Kernel hardness*:** The endosperm texture of the grain determines the hardness of wheat. It could be "hard" or "soft." The hardness of a given cultivar of wheat is genetically controlled and is not directly correlated to the protein content of the kernel (Yamazaki and Donelson, 1983). The hardness is measured in terms of 'Pearling Index' which is defined as the percentage of material peeled off from a wheat sample of known weight in a laboratory pearler for a prescribed time in seconds. Pearling is a process that with the application of abrasion and friction removes effectively only the bran layers from the cereal grains allowing nutritious parts such as the aleuran layer to remain in the intact kernels. Hard and soft wheat have different pearling index. Particle Size Index (PSI), Single-Kernel Characterization System (SKCS) 4100, and the Near-Infrared Reflectance Spectroscopy (NIRS) are the approved methods currently in use to measure grain hardness (AACC, 2000). Most of the wheat varieties developed in India are hard to very hard in the texture of grains as measured by SKCS method. The hardness index varies from 40 to more than 100 with average values of 82. Hard grains and medium strong gluten are preferred for good chapatti-making quality (Ram and Mishra, 2010). Based on SKCS approved method, Sharma et al (2012) have proposed the classification of Indian wheat types (*aestivum* and durum) according to kernel texture into **5** different classes (**Table 4**):

Table 4: Classification of Indian wheat types according to kernel texture.

Class	SKCS Score (Hardness Index)
Very soft	34 or less
Soft	35 to 54
Medium hard	55 to 74
Hard	75 to 89
Very hard	90 or more

Source: Sharma et al (2012). * Includes both released cultivars and genetic sources.

Grain hardness or kernel hardness affects parameters such as milling yield, starch damage and baking properties. Soft wheat has soft endosperm texture, requires less energy for milling and yields smaller particles with less starch damage upon milling than do hard wheats (Bass, 1988). Hard wheat upon milling produces flour with high levels of starch damage desirable for leavened and unleavened products, and grinds to give relatively angular particles that flow easily and are easy to handle. Soft wheat, on the other hand, mills to give flour with relatively small irregular-shaped particles that do not flow easily and tend to block sieves in the mill. These are generally used for cake, cookie and biscuit production because of their fine particle size and low starch damage. Medium hard to hard wheats are reported better suited than soft wheats for chapatti making (Austin and Ram, 1971).

- ***Colour*:** The wheat may be red, amber, white depending upon the colour of bran. Red wheats are dark in colour, tend to be harder and have high protein content. Red wheats are particularly suitable for bread making than the yellowish coloured wheats. Yellow pigment as ß-carotene in grains contributes to the desirable creamish whole meal appearance to the whole-wheat flour. Chapatti consumers preferentially choose varieties with bold, sound, uniform and lustrous amber golden colour grains which yield flour with creamish tinge and the resultant dough does not turn brown when kept for longer period, and maintains its consistency (Misra, 1998).
- ***Damaged wheat*:** Wheat may get damaged in field before harvesting due to bad weather, frost, heat, insect and mould attack, sprouting, etc. Damage may also occur due to thrashing and drying operations, handling, transport and storage. During storage excessive moisture left in the grains leads to fungal growth and discolouration of the germ, and the insect attack affects the quality of the flour upon milling. Any kind of damage to wheat not only affects the flour yield but also the quality of flour and consequently the quality of the product.

  The physical quality characteristics of wheat may be determined according to the methods as described by the American Association of Cereal Chemists (2000).

- ***Milling/grinding test*:** This test is conducted to assess the quality of flour for end use purposes. This test is exploited to screen the wheat varieties and to eliminate the varieties with unsatisfactory milling / grinding properties such as yield, ash content and colour grade. Murty and Austin (1963) conducted this test using a hand driven stone grinder called *chakki*. The ground flour was sieved on a 36 mesh sieve and the percentage yield of flour was calculated on the basis of the unsieved flour. They reported that soft wheats compared to hard and medium hard wheats tend to give low yield of flour. The lower yield is attributed to the production of fine flour of soft and amorphous particles, which do not flow freely through the sieve. The varieties that give higher yield of flour, on the other hand, produce flours of granular texture, which flow freely through the sieve. The higher flour extraction reported in the range of 93 to 95 percent in this study is, however, commonly used for chapatti making. The fine, soft, light and amorphous flours are more often used to make biscuits, cakes, etc. whereas for chapatti making flours of granular texture are preferred.

Chemical Quality

- ***Moisture*:** Moisture content of wheat is regarded as one of the most important quality characteristics, mainly because it directly affects the specific weight and value of wheat and its effect on the microbiological and germination stability of wheat during storage. Hence, moisture content of wheat grains is considered very important in grain trading and from storage point of view. The optimum storage moisture for wheat is < 12.5 per cent whereas optimum roller flour milling moisture is between 14 and 17 per cent, necessitating the addition of water (conditioning) to the wheat before milling. However, the traditional *chakki* mills seldom condition wheat for moisture.

  Moisture can be determined by heating in an air oven at 130°C for one hour or by using other instrumental techniques (AOAC, 2000).

- ***Ash*:** The ash represents the mineral content of a food material and has nutritional significance. The bran coat of wheat grain is the store house of minerals containing as much as 20 times more than the endosperm (Sramkova et al, 2009). The ash content is an indicator of milling efficiency and reflects the purity of wheat flour. The whole wheat flour of about 95% extraction, meant for chapatti making, yields higher amount of ash compared to wheat flour for bread. The ash content, however, results in undesirable darkening of dough and eventually discolouraton of chapatti. Also, the higher ash content of flour adversely affects the dough sheeting properties, machinability, and colour, taste, texture and pliability of chapatti (Parbhasanker et al, 2002; Panghal et al, 2019). The ash content is determined in a muffle furnace by incinerating the sample at 550-600°C.

- ***Crude fibre*:** The crude fibre content of wheat varies from 2.0 to 2.7 per cent (Austin and Ram, 1971). It is determined by digesting the sample with acid followed by alkali and then weighing the residue after drying. Wheat grain fibre has dietary significance.

- ***Lipids*:** Wheat contains about 1.5 to 3.0 per cent total lipids. Despite this low percentage, the importance of wheat lipids is recognized because of their nutritional value, as a source of fat soluble vitamins and polyunsaturated fatty acids, and their role in ageing of wheat flour, mixing behavior of dough, baking and freshness retention in baked products. Both polar and non-polar lipids play a role in determining the texture of chapatti. The wheat varieties having higher concentration of non-polar lipids adversely affect the texture of chapattis.

On the other hand, wheat varieties containing higher percentage of polar lipids are reported to give chapattis softer in texture (Arya et al, 1978).

- ***Protein*:** The primary function of protein in cereals is to provide nutrition. Besides, both quantity and quality of protein reflect the chapatti making quality of a wheat variety. The wheat varieties containing protein in the range of 10 to 12% are considered ideal for making chapattis of acceptable quality (Austin, 1978a). The quantity and quality of gluten, which is formed as cohesive mass upon mixing flour with water, play greater role in chapatti quality. However, very high content of protein and gluten of high strength makes the chapatti tough and leathery. On the other hand, wheat with very low protein (< 9 %) and weak gluten makes crumbly type of chapatti having poor texture and keeping quality (Austin and Ram, 1971).

  The protein content is generally determined by Kjeldahl method, although Dye Binding techniques and Near Infrared methods have also been used (AOAC, 2000).

Tests to Predict the Chapatti Making Quality of Wheat Varieties:

- ***Sedimentation value*:** This can be obtained from Zeleny's Sedimentation Test (Zeleny, 1947). This test is based on the principle that gluten protein in wheat absorbs water and swells considerably when treated with lactic acid. The degree of sedimentation of flour is taken as a measure of the baking quality. The amount of water absorbed and consequently the extent of swelling depends on the quality and quantity of gluten in the wheat. Both higher gluten content and a better gluten quality give rise to slower sedimentation and higher Zeleny test values. Muller (1970, 1973), however, reported that the sedimentation value is related to the granularity of the flour, and that sedimentation is an agglomeration of the coarse particles rather than swollen protein.

In this test wheat is coarsely ground and sifted to yield white flour which is suspended in water and treated with lactic acid. The volume of the sediment is measured in milliliter after a fixed time and is the indicator of the quality of gluten in wheat. Based on sedimentation values, wheats have been classified into four utility groups (Austin and Ram 1971):

Sedimentation Value	Gluten strength	Utility group
< 20 ml	Weak	Suitable for biscuits, cake. pastry, etc.
20 to 40 ml	Medium	All- purpose flour
40 to 59 ml	Strong	Bread flour
> 60 ml	Very strong	Best suited for blending purpose

The wheat varieties that were found to have good chapatti making properties showed sedimentation values varying from 30 to 37 ml and possessed more than 60 per cent water absorption capacity. The All India Coordinated Wheat and Barley Project, on the other hand, gives the sedimentation values of >60 ml, 30 – 60 ml and < 30 ml respectively for making good quality bread, chapatti and biscuits (DWR Progress Report, 2008). The wheat varieties having sedimentation values < 30 ml had water absorption capacity of < 60 per cent. The chapattis made from these varieties, in general, had a semi-stiff or stiff texture and had poor keeping quality as they tended to dry up quickly (Austin and Ram, 1971). It may, therefore, be emphasized that water absorption capacity of the flour is one of the important characteristics for determining the softness and pliability of the chapatti.

Sodium dodecyl sulphate (SDS) sedimentation value (SV) test (Axford et al, 1979) has also been used in place of Zeleny test to determine the sedimentation value of whole meal or wheat flour. The high SDS sedimentation values have been reported to be linked with stronger gluten and superior bread making quality and pasta quality, and low sedimentation values with soft wheat products making quality. Srivastava et al (2003) while evaluating the chapatti-making quality of 15 wheat varieties concluded that varieties having a protein content of about 13 per cent, SDS-SV of around 75 ml, and having 5+10 subunits at the Glu-1D chromosome, yielded excellent chapattis. Sharma et al (2004) reported that wheats with a range of 43 to 84 ml SDS-SV are suitable for chapatti making.

- ***Pelshenke value*:** The chapatti making quality of wheat is essentially related to the strength of gluten developed when whole-wheat flour is mixed with water. Pelshenke test (Pelshenke, 1933) is the measure of the strength of gluten and is extensively used in wheat quality improvement programmes. This test comprises kneading 3.0 g of whole wheat meal (*atta*) with 1.9 ml of 2.5 per cent yeast suspension into a small dough ball and observing its disintegration time in minutes due to fermentation in water maintained at 30^0C. The length of time required to beginning the disintegration of the dough ball is noted. This depends on the strength of gluten which provides the dough a particular form and consistency. The time required for wheat with weak gluten is less than 80 minutes while for strong/hard wheat it is more than 200 minutes. Based on these values, Austin and Ram (1971) classified wheat varieties tested into four broad groups:

Pelshenke value (min)	Quality of wheat Gluten
< 80	Weak
80 to 150	Medium strong
150 to 200	Strong
> 200	Very strong

Soft wheats belong to the weak group of wheats with Pekshenke values < 80 min. The flour of these wheats has water absorption capacity < 60 per cent. The dough handling property is poor, and it is difficult to roll and flatten the dough into circular disc because of stickiness. Puffing of chapatti, in general, is partial. The chapattis also assume a semi-stiff to stiff texture with poor keeping quality.

The flour from medium strong group of wheats with Pelshenke values between 80 and 150 min requires more than 60 per cent water to make dough of moderately stiff consistency. The dough is strong, stretchable, elastic and non-sticky. Generally, the chapattis puff rapidly and fully and possess a soft, smooth, silky and pliable texture. The desirable characteristics of the chapatti are retained for a longer time resulting in good keeping quality. The most desirable wheat varieties in respect of chapatti quality are, however, those that have Pelshenke values between 120 and 150 minutes.

The flour from strong group of wheats with Pelshenke values between 150 and 200 min also exhibits higher water absorption capacity and tends to make strong dough. These varieties also, in general, would make chapattis of good quality especially those varieties of wheat having Pelshenke values below 165 or 170 minutes. Higher Pelshenke value wheats have been found to make chapattis leathery particularly after keeping for over one hour.

The flour from very strong group of wheats with Pelshenke values above 200 min absorbs larger amount of water to make dough of required consistency. The dough is very strong and the chapattis, especially freshly made, are of satisfactory quality in general, but are felt tough and leathery while eating particularly when kept for over one hour after baking. The wheat varieties belonging to this group, however, may prove fit for blending purposes and for making bread.

- ***Biochemical Tests*:** Parbhasankar (2002) developed a rapid, sensitive method based on electrophoretic and immunological techniques to assess the chapatti-making quality of Indian wheat varieties. Sodium dodecyl sulphate-polyacrilamide gel electrophoresis (SDS-PAGE), and dot-blot assay using anti-gliadin antibody were used for predicting the chapatti-making quality of wheat. The wheat varieties with good chapatti-making quality showed distinct pattern in SDS-PAGE, and less intense spots in dot-blot assay. The results suggested that either SDS-PAGE or dot-blot techniques could be conveniently used for identification of wheat varieties suitable for chapatti making.

Similarly, Prabhasankar and Sai Manohar (2002) developed an indirect antibody-based enzyme-linked immunosorbent assay (ELISA) for prediction of chapatti-making quality of wheat varieties. Polyclonal antibodies against gliadin, low molecular weight glutenin (LMG), and high molecular weight glutenin (HMG) were developed and utilized in the ELISA. Out of these antibodies, an antigliadin antibody response was negatively correlated with farinograph water absorption, chapatti dough water absorption, overall chapatti sensory score, chapatti puffed height score, and positively correlated with shear value of chapatti. Anti-LMG antibody response was not correlated with any of these parameters. Anti-HMG response was, on the other hand, positively correlated with chapatti dough water absorption, farinograph water absorption, and overall chapatti sensory score, and negatively correlated with shear value and chapatti puffed height. Anti-gliadin antibody was, however, found to be more suitable in ELISA for evaluation of chapatti-making quality of Indian wheat varieties. The ELISA is simple and easy to use and is applicable to different wheat varieties. Wheat varieties with good chapatti-making quality showed less anti-gliadin antibody response. Among the 28 Indian wheat varieties tested for chapatti-making quality, NIAW 421, NIAW 34, PBW 343, PBW 373, HD 2189, GW 190, and LOK-1 had excellent chapatti-making quality, while six other varieties (HD 2501, NIAW 579, NIAW 301, HD 2735, DWR 235 and HI 977) had good chapatti-making quality. Nadeem et al (2015) applied the ELISA test to assess the chapatti making quality of different Pakistani wheat varieties. They, however, recommended that the use of antibody response against glutenin and HMW-GS offers a good method to predict the quality and suitability of different Pakistani wheat varieties for chapatti-making quality.

These biochemical tests are more sensitive and rapid and require minimal amount of wheat sample, and hence may be considered useful for wheat growers, bakers/ chapatti and related products makers, and the consumers.

- ***Fat acidity*:** Fat acidity refers to the breakdown of fats by lipases and the release of free fatty acids in the grain, which influence wheat quality. Under most practical storage conditions, fat acidity in wheat increases after several years to levels considerably above those associated with freshly harvested sound wheat, even though the wheat shows no appreciable physical evidence of deterioration. Such wheat may still be useful for milling purposes but the keeping and baking quality of the flour may be adversely affected. Low temperature and moisture content during storage markedly reduce the rate of increase in fat acidity. To ascertain the soundness of wheat grains, fat acidity test is performed. It is the number of milligrams of potassium hydroxide required to neutralize free fatty acids in 100 g of grains (calculated on moisture free basis). A value of fewer than 20 refers to the soundness of grains, whereas high damaged grains indicate values above 100 (Pomeranz, 1988).
- **Alpha amylase content:** Excessive high levels of alpha amylase can impair the quality of the dough and the final quality of baked products because of the rapid deterioration of the starch molecules leading to reduction in viscosity of the dough (Rasper and Walker, 2000).The alpha amylase test is performed to determine the sprout damage of grains especially in wet harvest conditions because the level of alpha amylase enzyme increases upon sprouting. Even if there is no visible sprouting, the prolonged wet conditions before harvest may raise the level of alpha amylase in the grains.
- **Polyphenolase Activity:** Compared to Indian tall wheats, the dough prepared from the whole wheat meal of some dwarf, high yielding Mexican varieties turns dark on resting at room temperature for1 to 3 hour; longer the resting period darker the colour of the dough, which subsequently gives darkish brown chapattis (Abrol, 1972). The wheat grains when treated with phenol solution may assume light brown, medium brownish black or dark brownish black colour depending upon the level of polyphenolase activity. The differences in discolouration are due to wheat varietal differences (Singh and Sheoran, 1972)

as depicted in the **Table 5.**

Table 5: Polyphenolase (Tyrosinase) activity of some Indian and Mexican wheat varieties

Variety	Phenol colour reaction	Enzyme activity (units/ml)
C 306	Slight change in colour	12
C 281	Slight change in colour	12
Sharbati Sonora	Dark Brown	22
Sonalika	Dark brown	23
UP 301	Black	29
Triple dwarf Lal Bahadur	Black	30

Source: Singh and Sheoran, 1972.

The change in colour negatively correlates with the chapatti making quality of wheat. Since different varieties of wheat develop different degree of darkness because of different levels of polyphenolase activity, this test may be used in identifying mixtures of wheat varieties.

Phenol colour test: In this phenol colour test about 100 wheat grains are taken in a glass test tube and one per cent solution of phenol is added to submerge the grains. After keeping for two hours, the solution is drained off and the grains are dried on a filter paper. The change in colour is recorded. The wheat variety that gives low response to phenol test possesses low polyphenolase activity and is considered of good chapatti making quality (DWR Progress Report, 2008).

The indigenous wheat varieties are reported to be low in these enzymes, whereas the imported Mexican hybrid wheat varieties still grown in India have a high content of these enzymes. The adverse effect of enzymatic browning is greater when the chapatti dough is rested for a longer period, and also when the baked chapatti is stored for a few hours. Besides the polyphenolase activity, sugars, amino acids and phenolics occurring in wheat grains and their increase on dough resting due to the activity of the enzymes *viz.* amylases, proteases and glycosidases, also contribute to the discolouration of whole meal dough and chapattis through non-enzymatic browning reactions and caramelisation (Abrol et al, 1972b).

Chapatti Flour or Whole-wheat Flour

Whole wheat flour is a coarse textured flour ground from the entire wheat kernel. It contains not only the endosperm but also the entire bran and the germ. It is rich in B-complex vitamins, vitamin E, fat, protein, and contains more trace minerals and dietary fibre than does white flour or all-purpose flour. The presence of bran, however, reduces gluten strength. The baked products made from it tend to be heavier and denser than those made from white flour. Flour quality is the ability of the flour to produce uniformly an attractive end-product, i.e. chapatti at a competitive cost under conditions imposed by the chapatti maker and the consumer (Austin and Ram, 1971).

The wheat variety that meets the above criteria of physical and chemical quality may be selected to mill using a suitable milling system to produce whole wheat flour of desired quality.

Milling of Whole Wheat Flour (*Atta*)

The devices to grind wheat into whole wheat flour (*atta*) for chapatti making have evolved from simple hand operated stone *chakki* to modern sophisticated roller flour mills. Traditionally, *atta* has been produced by stone grinding using manually (hand), water (*Gharat*) and power driven stone grinders commonly called '*chakkies*' These *chakkies* have a pair of circular granite stone discs/plates of varying sizes with chiseled inner surface and mounted horizontally one above the other. The system is provided with an opening at the top fitted with a suitable size hopper for grain entry, and a spout at the bottom to receive *atta*. In the manual *chakki*, the upper stone disc is rotated with the help of a wooden handle fixed upright at the top close to the outer edge of the disc. Grains are dropped gradually through the top opening in the centre of the upper disc and *atta* comes out through the gap between the two discs and collects in the rimmed periphery. Since the gap between the two plates is fixed, the granularity of flour may be adjusted by regulating the feed.

The manually driven *chakki* is a native system with a grinding capacity of only 2 to 3 kg per hour, and is used in villages, usually operated by women folk for preparing the *atta* for the day to day use. Electrically operated *atta chakkis* have now replaced the manual system. The water driven *chakki* (*Gharat*) is usually found in sub-mountainous regions where water from irrigation channels is directed to a steep decline to rotate the upper stone disc of the grinder which may weigh 400 to 500 kg. The upper stone disc rotates at an RPM of about 120 with an output of 20 to 30 kg per hour. The electrically powered stone mills on the other hand, operate at much higher speed than the water-driven mills consequently heat is generated due to friction between the chiseled stone surfaces during grinding. In these systems the upper stone disc weighs 500 to 600 kg and rotates at a much higher RPM with an output of about 300 kg per hour (Sharma and Bains, !976). Since there is a possibility of stone dust mixing with the flour during grinding, stone plates are being replaced with chilled cast iron corrugated plates mounted vertically having a grinding capacity of about 100 kg per hour.. One of the plates rotates at about 1500 RPM while the other is stationary. The wheat entering from the hopper at the top in between the plates is ground between the plates due to friction and shearing action and flour emerges through the outlet located at the bottom of the plates. The granularity of flour in either case can be adjusted by varying the clearance between the plates by adjusting the rotating plate, and the feed rate. Coarse particles of bran and endosperm may be removed with the help of vibrating screens of suitable mesh size.

Inamdar and Prabhasankar (2017) attempted to produce *atta* by varying the settings of the stone *chakki* in order to study the grinding characteristics of wheat, energy efficiency, characteristics of *atta* and its suitability for preparing chapattis of better quality and improved storability. Six settings (1 to 6 positions) were used by altering the gap (1.7, 1.2, 1.0, 0.7, 0.4 and 0.2 mm) between the pair of stone discs (*chakkis*). The *atta* produced by grinding at position four with gap of 0.7 mm was found to be of optimum quality in respect of particle size (about 81 per cent less than 300 μm), damaged starch about 18 per cent and water absorption about 80 per cent. The *atta* with these characteristics yielded chapattis with better sensory characteristics, texture and storage parameters.

The use of hammer, disc, pin and roller mills has also been studied to produce whole wheat flour for chapatti (Rao, 1982; Leelavathi et al, 1986; Rao et al, 1986b; Prabhasankar and Rao, 2001; Inamdar et al, 2015). However, there are differences in the quality characteristics of the flours. According to Leelavathi et al (1986) the flour samples milled in disc, stone and pin mill had significantly higher damaged starch (12-18.8%) and diastatic activity (365-505 mg/10g flour) as compared to hammer and roller flour milled flours (200-220 mg/10g flour). The dough water absorption varied from 68- 82 per cent with considerable variations in dough development time (3- 5.5 min) and dough stability (2 - 4.5min) as indicated by the farinograms, and resistance to extension (410 -590BU) and extensibility (50–65mm) as indicated by the extensogram (**Table 6**). Overall, the chapatti-making quality of *atta* produced in stone *chakkis* is superior as chapattis prepared from it are highly rated sensorially because of the highly desirable wheaty aroma and sweetish taste, better texture and overall eating quality. On the other hand, those chapattis made from hammer milled or roller milled flour were inferior with somewhat hard texture and bland flavour. *Chakki atta* has become the favourite of chapatti consumer world over.

Table 6: Effect of grinding wheat in different mills on the physicochemical properties of whole wheat flour.

Type of mill	Temperature (°C)	Moisture	Diastatic activity (mg/10g)	Damaged Starch (%)	Iron (mg %)	Water absorption (%)
Disc (*chakki*)	78	7.8	270	12.9	16.4	77
Pin	61	8.8	365	12 0	11.8	74
Hammer	50	9.5	220	9.9	10.3	70
Stone	92	7.4	505	18,8	7.5	82
Roller	31	9.9	200	8.2	10.1	68

Source: Leelavathi et al (1986).

Physicochemical and Functional Properties of Whole Wheat Flour (*Atta*)

The functional properties of whole wheat flour depend on several factors: wheat variety, environmental and soil conditions, and agricultural practices under which the wheat was grown, the process and device used to mill the wheat into flour, flour extraction rate, and the chemical composition of the flour (Austin and Ram, 1971). The chemical composition of whole wheat flour varies widely depending largely on the wheat variety, and the growing conditions. Some of the properties of whole wheat flour that influence the quality of chapatti are discussed below:

Flour Granulation / Particle size profile

Depending upon the type of mill used and the milling conditions, the wheat flour contains a range of particle size. It is well recognized that the particle size has great influence on the physical and chemical properties of the flour (Pomeranz, 1988). The flour granulation and severity of grinding play important roles in influencing the dough characteristics and the quality of chapatti (Sharma and Bains, 1976; Rao, 1982). The particle size profile of flour considerably influences the water uptake during dough making. The flours with smaller average particle size exhibit higher absorption values because of greater specific surface, and higher damaged starch content (Rao, 1982).

Aziz and Bhatti (1960) reported a wide range of granularity of the Pakistani bread wheat whole meals (*atta*) milled in various ways (hand operated stone *chakki*, roller mill, etc.). They found that flour should be fine, about 90 per cent passing through 50 mesh sieves; if less passed this sieve, chapatti quality will be deteriorated. Chaudhry (1968) experimented with US wheats for their suitability for making chapattis. The samples were milled in the Ross Experimental roller mill to produce *atta* of 80 and 95 per cent extraction. The particle size distribution of flours determined by the Rotap method was considerably different for different wheats but within the limits of variation of Pakistani samples. This was unlike the customary grinding of wheat into 100 per cent extraction *atta* in power driven stone *Chakkis* where unconditioned wheats are used. The particle size stipulation according to this author, however, remained elusive in relation to chapatti quality with preference for finer flours. Sharma and Bains (1976) used hand, water and power driven *chakkies* to compare the granularity of whole wheat flour determined in a Rotap machine, and reported that hand driven *chakki* produced significantly coarser flour than those from water and power driven *chakkies*. The overs on the 355μ sieve were: 37.0, 12.7 and 4.7 per cent of flours obtained from hand, power and water driven *chakkies*, respectively. The water driven *chakki* yielded relatively fine flour as evident from the higher percentage of overs (72.6 per cent) on 150μ sieve compared with 25.2 and 34.4 per cent of the whole wheat flours obtained in hand and power driven *chakkies*, respectively. Thus the whole wheat flour obtained from water driven *chakki* was found superior for making chapattis of good quality. Rao (1982) used disc-milled whole wheat flour and reported that very coarse and medium coarse flours containing 77 and 67 percent over tailings in 129μ sieve respectively yielded chapattis with tough texture and poor overall quality, while chapattis made from normal or fine flours containing 31 to 54 per cent over tailings in 129μ sieve were of better quality with high acceptability. Leelavathi *et al* (1986) investigated the functional characteristics of whole wheat flour milled in stone, disc, hammer, pin and roller mills. Sieve analysis of the flours obtained from different mills was carried out in a Buhler laboratory Plansifter showing retention of less than 7 percent of *atta* over the 340μ sieve in most cases. Maximum amount of *atta* (24.5 to 38 per cent) was left over 112μ sieve in almost all the cases. Similar observations were made by Rao *et al* (1986b) for *atta* milled in a hammer mill and a disc mill, and found that the particle size distribution of flours ground in different disc mills varied widely for lack of provision to control the gap between the two discs. On the other hand, Islam and Johansen (1987) who used hammer mill reported that whole wheat flour of particle size of 1000μ and below was satisfactory for making chapattis with desirable sensory characteristics. Sidhu *et al* (1988) studied the effect of different milling conditions on the granularity of roller mill whole wheat flour and concluded that finer flours having more of flour particles below 250μ produced chapattis of highly desirable texture. It was also shown that inclusion of fine bran did not adversely affect chapatti texture, but the same amount of coarse bran had severe detrimental effects. These findings have been corroborated by Rao (1993) that the fine-ground flour is desirable for chapatti making and the coarse flour with 70-80 per cent and more over tailings on a 129μ sieve makes chapatti having a poor appearance and tough texture. From the above observations it may be concluded that stone milled *atta* makes superior quality chapatti than those obtained from other mills.

Since a number of grinding devices have been used to grind different varieties of wheats, there is likelihood of vide variation in the granulation or particle size distribution of whole wheat flour. Particle size may be determined in Buhler Plansifter, and using the Rotap assembly of sieves of known aperture diameter, benzene sedimentation method, air classification technique and other modern instruments (AACC, 2000) for standardization.

Flour Extraction Rate

Besides the particle size of whole wheat flour, flour extraction rate too affects the final quality of the chapatti. The term flour extraction is simply and generally defined as the proportion of flour by weight, derived by milling, from a known quantity of wheat. Traditionally, 100 percent extraction whole wheat meal is used to make chapattis with some households, however, sieving off 4 to 6 percent coarse bran before making dough, thus virtually using 94 to 96 percent extraction flour. This practice though does not affect the quality of chapatti much, removing higher amounts of bran by sieving, or using a flour of a lower extraction rate makes the chapatti tough and leathery (Rao 1993). However, in urban areas, mechanical abrasion of wheat prior to milling is commonly practiced, and variable percentage of coarse bran and endosperm are sieved off at the mill itself resulting in very fine whitish whole wheat flour of 80 to 86 per cent extraction, which the urban consumers prefer as it yields more whitish chapattis (Kaur *et al*, 1990).It has been shown that the abrasive removal of bran from wheat prior to milling results in not only considerable losses of essential nutrients (protein, minerals, fat, crude fibre, essential amino acids, etc.) but also affects adversely the textural characteristics and flavour of chapattis (Kaur *et al*, 1990; Bajaj *et al*, 1990).It is widely acclaimed that the chapattis made from white (75 percent extraction) flour are "bad for the digestion" (Chaudhri and Muller, 1970). Hatcher *et al* (1997) evaluated a wide range of extraction rate roller mill flours of Canadian wheats for their chapatti making quality and concluded that low-extraction flours were less suitable for chapatti making. The 85 per cent extraction flours, however, yielded the preferred chapatti having creamy yellow colour, wheaty aroma and taste, and soft, flexible texture with easy tearing.

Whole wheat flour temperature

The frictional grinding of wheat in stone or disc mills results in *atta* attaining high temperatures up to 95°C. Inamdar et al (2015) recently reported variations in flour temperature from *chakki* (75- 84°C), disc (69- 78°C), and hammer mill (67-74°C) compared to roller milled flour (35-38°C). The heat thus produced is reported to have both positive and negative impact on the physicochemical status of the flour (**Table 7**). There is decrease in moisture content and total microbial load, which help increasing the storage life of the flour. The quality of chapatti particularly with respect to flavour and taste is reported to enhance because of the increased diastatic activity owing to higher levels of damaged starch (Ranga Rao et al, 1986). Earlier Rao (1982) made similar observations and concluded that heat developed during grinding of wheat into *atta* had beneficial effects on the overall quality of chapatti. However, high flour temperature has a significant effect on bound lipids, protein fractions, and diastatic and proteolytic activities of the flour (Ranga Rao et al, 1986). There is partial destruction of vitamins thiamin and niacin (Austin and Ram, 1971). Prabhasankar and Rao (2001) also observed that heat produced in stone and plate mills denatures the protein, and reduces the amino acid and the unsaturated fatty acids content of *atta*.

Table 7: Effect of heat developed during grinding (disc mill) on some characteristics of whole wheat flour.

Whole wheat flour	Moisture	Lipids		Proteins (N x 5.7)				Damaged starch	Enzyme activity			Total plate count
		Free	Bound	Total	Salt soluble	Gliadin	Glutenin		Diastase	Lipase	Protease	
	(%)	(%)	(%)	(%)	(%)	(%)	(%)	(%)	(mg/10g)	(ml NaOH)	(mg /100g)	(x 10^3/g)
Control*	10.2	1.40	0.99	11.6	2.5	4.2	4.9	4.5	196	4.62	30.5	15.3
Cooled immediately	6.8	1.74	0.73	11.5	2.8	4.5	4.2	12.1	373	4.54	27.1	3.9
Cooled slowly	7.2	1.74	0.72	11.4	2.7	4.5	4.2	12.1	367	4.65	27.0	1.5

Source: Ranga Rao et al (1986). *Ground in pestle and mortar.

Whole wheat flour colour

Indian consumers inadvertently prefer lighter coloured *atta*. The most desired colour is creamish white. It should not darken during storage, and should yield creamish dough which does not darken upon storage. It is the amber coloured wheat grains that yield creamish coloured flour (Misra, 1998).

Flour moisture

Freshly milled *atta* may contain moisture somewhere between 7 and 10 percent depending upon the moisture content of the wheat and the type of mill used (Rao, 1993). There is loss of moisture when wheat is milled in stone or disc mill as due to severity of grinding *atta* gets heated to a temperature up to 95^0 C by the time it comes out of the mill (Ranga Rao *et al,* 1986). Rise in moisture content during storage reduces the storage life of flour causing ropiness and rancidity. The rheological properties of *atta* change considerably with period of storage. Water absorption, dough development time, valorimeter value and resistance to extension are reported to increase, while extensibility and weakening of dough upon mixing decrease (Arya et al, 1971).Moreover, the chapatti baking quality of the *atta* is adversely affected upon prolonged storage.

Flour Protein

Protein quality and composition, wet and dry gluten content in whole wheat flour play an important role in overall acceptable chapatti quality (Sharma et al, 2004). Gluten is the cohesive mass derived from the protein when flour is mixed with water. Gluten is itself a mixture of three constituents, namely gliadin, glutenin and residue proteins (Ram and Nigam, 1979). Gliadin is very extensible but has poor flexibility; with water it yields long but thin ribbons, and contributes mainly to the viscosity and extensibility of the dough. The glutenin, on the other hand, is of opposite characteristics i.e., low extensibility, highly cohesive and elastic, and is responsible for dough strength and elasticity (Macritchie, 1984). With water glutenin forms thick sheet which is quite flexible but will break when pulled apart. Though glutenins from different wheats themselves vary quite a bit in their properties, they all contribute strength to wheat flour dough. Glutenin forms complex with lipids. The third gluten component the residue proteins which are unextractable from gluten in the presence of acetic acid contribute significantly to dough extensibility. The relative quantities of these three gluten components play an important role in rheological properties of dough and in determining the chapatti characteristics. During dough formation the extensibility and elasticity is provided by gliadin and glutenin, respectively. The balance between the two properties, though genetically controlled, can differ greatly between varieties and has a major influence on the type of food product that can be prepared. The wheat varieties, the flour dough of which have intermediate values for extensibility and elasticity, are considered good for chapatti making (Payne et al. 1987). Ram and Nigam (1979) fractionated whole meal flours from 11 wheat varieties of varying strength into different protein components and found intervarietal quantitative differences between these gluten components. When wheat varieties were arranged in the order of decreasing flour strength in terms of dough stability and dough development time, the ratio of residue protein to gliadin + glutenin was found to decrease (**Table 8**). The farinographic properties, dough stability and dough development time of these wheat varieties was found to positively correlate with this ratio. According to the authors, this ratio can be used as an index of flour strength, which may prove useful in formulating flours of desired strength for producing bakery products of desired quality.

Table 8: Composition of flour proteins of selected wheat varieties as percentage of total protein

Variety	Albumin A	Globulin B	Gliadin C	Glutenin D	Residue proteins E	DDT (min)	Dough stability (min)	E /C+D Ratio
K68	10.5	17.1	21.1	18.4	32.9	9.0	11.5	0.833
K65	12.5	16.6	22.3	19.4	29.2	8.0	9.0	0.700
Hyb65	14.7	10.4	24.0	26.1	24.8	5.0	6.0	0.495
HD4530	11.9	10.4	28.9	26.3	22.7	4.5	6.0	0.411
Kalyansona	17.9	7.5	26.6	28.1	19.8	4.5	4.5	0.362
HD1949	14.0	9.6	31.9	30.0	14.5	4.0	4.0	0.234
C306	16.0	9.8	35.4	27.3	11.4	4.0	3.0	0.182
UP K1	14.5	10.5	28.7	26.1	20.1	3.0	3.0	0.367
HD2189	16.0	11.1	31.9	32.0	9.1	2.5	2.3	0.142
UP283	13.6	11.5	32.7	33.8	8.4	2.5	2.0	0.126
Sonalika	12.9	16.4	34.2	31.7	4.8	2.5	1.3	0.073

Source: Ram and Nigam (1979). DDT= Dough development time

The studies conducted by Ram and Nigam (1981) have further shown that wheat varieties having roughly equal quantities of these three gluten proteins will make better chapattis than varieties in which excessive amounts of one of the other gluten component is present. The residue proteins appear to play a greater role in chapatti quality. Both low and high proportions of residue proteins are detrimental to chapatti quality. Low proportion of residue proteins is associated with stiffness and partial puffing and high proportion of residue proteins with leathery texture and full but slow puffing of chapatti. Prabhasankar (2002) reported that wheat varieties with low gliadin content were more suitable for making chapattis. Sharma et al (2004) have shown the possibility of the glutenin fraction of gluten being important to the chapatti making quality of whole-wheat flour. The glutenin fraction of gluten consists of two main protein subgroups: high molecular weight (HMW) and low molecular weight (LMW) subunits which occur in flour in proportions from about 1:2 to 1:3 (weight basis) depending on the wheat cultivar (Weiser et al, 1994). The molecular weight distribution of glutenins has been recognized as one of the main determinants of dough properties and baking performance. Since bread-making quality is highly influenced by the HMW subunits of glutenin (Payne *et al*, 1987), they may also play a role in determining chapatti making quality of Indian wheats as well. Bhatnagar et al (2002) tested five wheat varieties differing in chapatti quality characteristics viz. C306, K68, HD2745 and HD2735 with good, and Sonalika with poor chapatti quality characteristics for characterization of glutenin genes. The study suggested that HMW glutenin subunit Dx5 and Dy10 may be responsible for good chapatti characteristics of wheat varieties. Ram (2003) evaluated 172 wheat varieties including 25 durum wheat cultivars for high molecular weight glutenin subunit (HMW-GS) composition using sodium dodecyl sulphate-polyacrilamide gel electrophoresis (SDS-Page) and studied the relationship between HMW-GS and sedimentation value for dough strength. Protein subunits 13 + 16 and 6 + 8 were found correlated positively with improved dough strength as compared to subunit 20 in durum wheat. Srivastava *et al* (2003), while evaluating 15 Indian wheat cultivars differing in protein content and sodium dodecyl sulphate sedimentation value (SDS-SV), observed that excellent chapatti quality was associated with the presence of HMW glutenin subunits 5+10. The subunits 2+12, on the other hand, were found to be associated with poor chapatti quality. Das *et al* (2006), however, did not find significant differences in chapatti making quality of wheat selections differing in specific HMW subunits of glutenin indicating that individual HMW subunits / pairs are less likely to be a major factor in influencing chapatti making quality. Kaur et al (2013) evaluated 18 Indian wheat varieties differing in high molecular weight subunits (HMW-GS) for diversity in wheat grain, flour, dough and gluten properties. The varieties with HMW-GS of 20 and 2+12 showed very weak dough stability (DS), and varieties having 2+12 and 7+9 as well as 5+10 with 7 or 7+9 had weak DS, whereas those varieties with subunits of 17+18 with 2+12 or 5+10 as well as of 7+8 with 5+10 had strong DS. The grain hardness index (GHI) was negatively correlated with grain weight, flour recovery and proportion of fine flour particles, whereas gluten index (GI) was negatively correlated with wet and dry gluten content. The DS showed significant correlation with sedimentation value (SV) of both meal and flour, however negatively correlated with wet and dry gluten content. The water absorption was positively correlated with flour protein content and negatively correlated with SV. The flours with higher proportion of fine particles showed less dough development time and had higher DS. The elastic modulus and viscous modulus of gluten was positively correlated with quality number, SV and DS. It was concluded that GI, wet and dry gluten content should be cautiously used while evaluating the baking quality of Indian wheat varieties / cultivars. Katyal et al (2016) tested 28 Indian wheat varieties for diversity in quality traits and showed wide variation in both unextractable polymeric and monomeric proteins as well as various wheat flour properties. The majority of the varieties with HMW-GS combinations of 91 kDa+80kDa+78kDa+74kDa polymeric proteins showed very high grain hardness index (GHI). The varieties with higher GHI yielded flours with higher ash and protein contents, and greater flour colour redness. The unextractable polymeric proteins showed a positive correlation with gluten index (GI) and lactic acid solvent retention capacity. It was projected that GI may be useful in determining gluten strength. Goel et al (2015) have demonstrated for the first time a positive effect of wheat storage globulin triticin on the bread- and chapatti- making qualities of wheat. The authors suggested that the end-use quality of wheat varieties can be improved by combining specific alleles of triticin.

Flour Starch

Wheat grain may contain starch to the extent of 60 to 75 per cent of the grain dry matter (Sramkova et al, 2009). Since the whole wheat flour used for chapatti is of hundred per cent extraction, it should contain starch in this range. Starch is the main source of energy in the wheat based diets. However, the starch component in wheat flour imparts several functional properties (Pomerantz, 1988) to the baked goods, namely:

(a) dilution of gluten to make dough of desirable consistency,
(b) furnishes additional amount of sugar through amylolytic actions,
(c) provides a surface suitable for a strong bond with gluten,
(d) offers flexibility during the course of heating, and
(e) during complete gelatinization, starch takes water from the gluten to cause a firm set.

Flour sugar content

The sweetishness and flavour of chapatti is closely dependent on the amount of sugar originally present in the flour combined with the sugar produced through the hydrolysis of starch by the diastatic enzymes when the dough is made and kept for sometimes before chapattis are baked (Singh and Bailey, 1940).The whole wheat flour commonly used for chapatti making is reported to contain total sugars in the range of 2.5 to 4.0 percent (Rao, 1993). Extensive chapatti palatability tests have shown that this range of sugar is in fact needed to make chapatti sweetish and palatable to the ordinary chapatti eaters. The whole wheat flour that contains less than 2.5 per cent total sugar yields chapattis that score poorly for palatability and are more often judged insipid, and bland (Austin and Ram, 1971). On the contrary, the dwarf wheat varieties, e.g. *Lerma rojo,* though high in initial sugar content, yield chapattis of inferior quality because, dwarf wheats are relatively soft with lower values for starch damage during milling and consequent lower diastatic activity in the dough (Abrol and Mehdi, 1971). The amount of sugars produced in dough depends upon two factors, viz, proportion of amylolytic enzymes in flour and susceptibility of starch substrate to their attack. The latter factor is associated with the degree of starch damage caused by the milling process. Hard to medium hard wheats as compared to soft wheats yield more damaged starch upon milling because of severity of grinding required to yield flour of suitable particle size.

Flour damaged starch content

Starch in wheat occurs as granules. Damaged starch is that which have lost their crystallinity rendering starch granules susceptible to attack by starch degrading enzymes. During milling of wheat, part of the starch granules about 8 to 20 percent, are damaged i.e. reduced to smaller granules with more exposed surface. The extent of damage depends upon the type of grinding device used, and of course the type of wheat, hard or soft. Hard to medium hard wheats compared to soft wheats suffer greater damage to starch during milling because more pressure needs to be applied for milling such wheats (Bass, 1988). Farrand (1972) has related the extent of starch damage to protein content of wheat milled for bread flour. The optimum starch damage equals $(\text{protein})^2/6$ that is optimum starch damage is 20 percent at 11 percent protein and 24 percent at 12 percent protein in the flour subject to the method used to determine damaged starch.

Leelavathi *et al* (1986) milled the same variety of medium hard wheat into whole-wheat flour in different milling devices, *viz.* stone, disc, pin, hammer and roller mills and reported wide variation in damaged starch content of differently milled flour samples (**Table 6**). The highest amount was found in flour ground in stone mill (18.8 per cent) while the lowest was in roller mill flour (8.2 per cent). The flours from disc, pin and hammer mills, on the other hand, had a starch damage content of 12.9, 12.0 and 9.9 per cent, respectively.

The samples of whole-wheat flour milled in different types of stone mills, viz. hand driven, water driven and power driven *chakkis*, also exhibited considerable differences in damaged starch content largely due to differences in the severity of mechanical action of the grinders and the resultant particle size distribution (Sharma and Bains, 1976). The whole wheat flour prepared in water driven *chakki* (*gharat*) had the highest damaged starch contents determined by different methods. Recently, Ghodke *et al* (2009) have shown that the gap between the two grinding stones, feed rate and moisture content of grains control the particle size and damaged starch content of the resultant flour. These observations have been recently corroborated by Inamdar and Prabhasankar (2017).The milling conditions like narrower the gap, slower the feed rate and higher the moisture content, yield flour of higher damaged starch content (26.9 percent). These studies show that finer the flour, higher the content of starch damage with the exception of roller flour mill *atta*, which, though fine in granularity, contains least amount of damaged starch. This anomaly may be due to the mechanical action on the wheat grains during the process of milling, whereby the grains are subjected to fine cutting action of sequential reduction rollers in roller flour mill rather than attrition and abrasion action experienced by the grains in disc or stone mills.

The damaged starch content of whole-wheat flour markedly influences the quality of chapatti in terms of texture, taste and flavour. The damaged starch produced during milling has enhanced capacity to absorb more water. The farinograph water absorption may vary from one to two times of the dry weight of the damaged starch. Water uptakes of 0.44 and 2.0g per gram of native and damaged starch, respectively have been reported for bread flour (Greer and Stewart, 1959).The additional amount of water thus absorbed is not only conducive to the release of sugars by the enzymes but also gives flexible dough that permits the chapatti to be rolled really fine. The sugar released upon enzymatic breakdown imparts sweetness to chapatti. Starch damage is largely responsible for differences in water absorption, handling properties of dough, sugar production, and slackening of dough during resting. Starch damage also affects the crumb tenderness and puffing of chapatti. The higher the damaged starch content, the higher will be the water absorption and higher the starch gelatinization, and consequently, softer will be the texture, and longer the storage life of the chapatti (Sharma and Bains, 1976; Rao *et al*, 1986a: Rao *et al*, 1989). Rao *et al* (1989) investigated the effect of damaged starch in whole wheat flour on different properties of chapatti dough as well as chapatti, and reported an optimal proportion of 14 to 16.5 per cent damaged starch in whole-wheat flour milled in disc mill. Damaged starch in the flour was found positively correlated to diastatic activity and chapatti water absorption, and negatively correlated to per cent over-tailings on a 10 XX sieve. The various rheological characteristics of whole wheat flour were also influenced by damaged starch, as indicated by its significant correlation to dough development time, extensibility, resistance to extension, cohesiveness and adhesiveness. Flours containing damaged starch in this optimum proportion yielded chapattis with better texture, pliability, taste and overall acceptability. The chapattis made from flours containing higher proportion of damaged starch were found to remain fresh for over 4 hours as these retained more moisture because of the presence of more sugars and dextrins, which possess relatively greater water-holding capacity. According to Austin and Ram (1971), a good quality chapatti should retain its freshness for a minimum of 4 hours. However, the damaged starch in flour exceeding 16.5 per cent proved detrimental as the dough tended to become highly sticky and difficult to sheet or roll. The springiness of chapatti crumb is also impaired. Rehman et al (2007) studied different flour parameters of British wheat cultivars and reported that the damaged starch content was well correlated with the water absorption, chapatti texture, and sensory scores. Ghodke *et al* (2009) reported that dough stickiness increased with decreased gap between the grinding stones with consequential increase in damaged starch content, but softness of chapatti improved with decreased gap and increased moisture content of grains. Inamdar and Prabhasankar (2017) have recently reported that *atta* with damaged starch content of 18 per cent and water absorption of 80 per cent yielded chapattis with better acceptable sensory characteristics and storage properties.

The presence of damaged starch in flour is tested using the dye Congo red. The damaged starch granules stain red in this dye. Since the damaged starch becomes more susceptible to enzymatic action, this property is exploited to estimate the damaged starch content of flour. The method measures the net quantity of maltose released when flour is treated with the enzyme α-amylase, and damaged starch calculated by multiplying with the factor 1.64. The results are expressed as percentage of damaged starch in the flour (Austin and Ram, 1971).

Flour diastatic activity

The diastatic activity also called the "diastatic power" or "enzymatic power" of wheat flour refers to the flour's ability to breakdown starch into sugars. The diastatic enzymes, α- and β- amylases occurring in flour, act on starch when wheat flour is mixed with water to form dough of suitable consistency. Continued enzymic hydrolysis results in the formation of glucose, maltose and low molecular weight polysaccharides called dextrins with both amylose and amylopectin as substrates. These enzymes work efficiently on gelatinized starch. The native starch is least attacked by these enzymes. However, the enzymes work at a faster rate on damaged starch formed during milling of wheat into flour. Beta- amylase hydrolyses damaged starch only slowly. Alpha-amylase serves to condition the damaged or "available" starch so that it is more readily converted to maltose by beta-amylase. More severe the grinding, higher the damaged starch content, consequently higher the maltose value. The maltose values of whole wheat flour prepared in water driven *chakki* were the highest compared with those of the flours of hand and power driven *chakkies* (Sharma and Bains, 1976). The maltose value of flour is very important to both miller and baker or chapatti maker. It is the measure of diastatic capacity of the flour, i.e. the sugar producing power or maltose value. Higher the maltose value, better the chapatti quality in terms of texture, flavour, and pleasant sweetish taste (Austin and Ram, 1971; Rao *et al*, 1989). Sweetness and flavour of chapatti are due to the amount of sugars originally present in the flour combined with the sugars produced through the hydrolysis of starch by the diastatic enzymes when the dough is prepared and kept for sometimes before making chapattis (Singh and Bailey, 1940). It is, therefore, necessary that *atta* used for making chapattis should have desirable combination of total sugars and diastatic activity. Accordingly, the wheat varieties having suitable combination of these characters should be selected for chapatti-making. The varieties such as NP 824, NP 830, NP 880, NP 858, NP 860, HD 60-534, K 64, K 65, WH157, WH711, *Kalyan sona* and C306 were reported to have promising combination of total sugars and diastatic activity (Austin and Mirchandani, 1971; Popli and Dhindsa, 1980). Abrol and Mehdi (1971) found that tall wheats like C 273, C 591 and K 68 had higher diastatic activity when compared with dwarf wheats like Sonora68, *Lerma roja* and *Sonalika*. Differences in diastatic activity of flours from different types of wheat are attributable at least partly to differences in the physical hardness of the endosperm, for hardness affects the extent of starch damage during milling (Pomeranz, 1988). Quite high diastatic values, 250 - 500 mg maltose per 10 g flour and even higher values have been observed in the experimentally milled flours of Punjab wheats (Singh and Bailey, 1940) and in later studies of the milling and baking quality of several Indian wheats (Lyallpur Reports, 1943 - 46; Bains and Irvin, 1965; Sinha *et al*, 1968; Austin and Mirchandani, 1971; Austin and Ram, 1971; Abrol and Mehdi, 1971). Sinha et al (1968) reported that whole wheat flours with maltose number between 2.56 and 4.88g/100g are quite satisfactory for chapatti making. However, the excessive diastatic activity results in sticky dough and chapatti upon baking develops darker unappealing spots on the surface. The ill effects of wheats with too much diastatic activity arising due to wet harvest or very high starch damage upon severe milling action may best be overcome by blending with other wheats or flours of lower diastatic activity.

The alpha- amylase activity of flour may be determined by Hagberg Falling Number method or by amylograph (AACC, 2000). Falling number is an index of alpha-amylase activity of flour. The level of α-amylase in the flour increases when the wheat grains are exposed to humid and other undesirable climatic conditions during storage. Falling number affects the dough handling properties as lower falling number results in stickier dough (Prabhasanker et al, 2002). The diastatic activity of flour is estimated by measuring the amount of reducing sugars formed by the amylases in the flour under prescribed conditions. The amount of free sugars is subtracted from the total sugars derived from one hour diastasis of whole meals. The results are expressed as maltose value (MV) in gram maltose percent or mg maltose per 10 g flour (Austin and Ram, 1971).

Flour rheological properties

Rheology is the study of deformation and flow of matter in response to an applied stress and strain (Bourne, 2002).The bread baking and chapatti making quality of wheat flour (*maida*) or whole wheat flour (*atta*) is reflected in its performance in making the end products with desired physical, chemical, and sensory properties (Pomeranz, 1988). Dough, a mixture of flour and water, is an intermediate stage in the conversion of wheat flour to baked goods. Consequently, rheological properties of dough during mixing, rounding or ball making and moulding or sheeting / flattening become very important as they influence the quality of the end product. The importance of understanding the rheological properties of dough is enhanced further because of the mechanization and automation in chapatti production.

The wheat flour dough has unique visco-elastic properties (Schofield and Scott Blair, 1932).These two properties, viscosity and elasticity relate to the extensibility, elasticity, stability, strength, etc. of the dough and greatly influence the baking quality of baked goods. The ratio of viscosity to elastic modulus is considered of fundamental importance in determining baking quality. For the production of bread of good quality, dough should have high viscosity / modulus ratio. Since low-extraction flour (white wheat flour) and high-extraction flour (whole wheat flour) differ in their physicochemical properties, the rheological properties are likely to differ (Pomeranz, 1988). The preparation of chapatti dough of desired quality requires addition of optimum quantity of water and mixing for optimum time in a mixer or when kneaded by hand. The desired characteristics that determine the optimum quality of chapatti dough include optimum values for consistency, stability, elasticity, cohesiveness, extensibility, resistance to extension, adhesiveness as well as the strength of the dough (Rao, 1982). These chapatti dough characteristics are considerably influenced by:

- the cultivar and variety of wheat used,
- extraction rate of the whole wheat flour,
- particle size status of flour,
- amount of water,
- mixing time'
- type of mixer used, and
- level of optional ingredients like common salt and fat or other functional additives when used.

The chapatti dough characteristics such as elasticity, cohesiveness, extensibility, resistance to extension, adhesiveness determine the ease of rolling/sheeting, formation of continuous dough sheet without breaking, and the retention of the desired shape of the rolled dough disc before transfer to the baking griddle (Rao, 1982). The dough elasticity could be an important parameter in stabilizing the puff in chapatti during puffing act (Chakrabarti-Bell et al, 2013). The wheat variety that yields optimally elastic dough should, therefore, be selected to make chapattis that puff fully with highly acceptable quality. Uniform sheeting of dough is also considered important as the resulting thickness of the rolled raw chapatti affects the final quality of the baked chapatti. According to Rao et al (1986b), the optimal thickness of raw chapatti should be in the range of 2.0 to 2.5mm to get fully puffed, soft and pliable chapatti. The raw chapattis of 1.0 to 1.5mm thickness though puff fully the eating quality of the baked chapattis is reported to be inferior and unacceptable as these have coarse texture and dry up rapidly after baking. On the other hand, the raw chapattis with thicknesses greater than 2.5mm yield baked chapattis with under baked qualities. The sensory texture of chapatti is also related significantly to farinograph consistency, extensograph extensibility, resistance to extension, and texturometer cohesiveness (Rao, 1993). Manu and Prasad Rao (2008) have reported that total polymeric protein content in the flour protein showed a significant positive correlation with dough hardness as well as chapatti texture. Among the SDS extractable polymeric proteins, large polymeric protein in flour protein had significant positive correlation to dough hardness and chapatti cutting force, which reflects the chapatti texture. The flour protein disulfide content showed positive correlation to dough hardness and texture of chapatti while flour protein thiol content showed significant negative correlation to chapatti texture.

Several instruments have been developed to measure the rheological properties of the dough to predict the suitability of flour for various baked products including chapatti. The most important commercially available instruments include: recording dough mixers (Brabender farinograph, mixograph); load extension meters (Brabender extensograph, Chopin alveograph); extrusion meters (Research water absorption meter); viscometers (Hagberg falling number, Brabender visco-amylograph), texturometer, etc (AACC, 2000). The data generated from these instruments has revolutionized the bakery products processing industry since the entire processing is conducted on the basis of accurate measurement of dough behaviour rather than by using personal skill and judgments.

Austin and Ram (1971) used the alveograph to study the rheological properties (such as stability, strength, extensibility) of dough prepared from whole wheat flour (*atta*) derived from different wheat varieties to determine their chapatti making quality. The results showed that the values for stability vary from 63 to165 mm, for strength from 6.8 to 35.6 cm^2 and for extensibility from 18 to 35 mm among different varieties tested. From the stand point of appearance, texture and keeping quality of chapatti, whole meal dough of the wheat varieties having stability values varying from 100 to 130 mm combined with values from 15 to 23 cm^2 for strength and from 20 to 27 mm for extensibility make ideal chapattis. Some of the important varieties that yielded chapattis of very good quality included NP 824, NP 830, NP 871, C 306, and RS 31.1.The wheat varieties that showed values of over 140 mm and 23 cm^2 for dough stability and strength respectively result in chapattis having undesirable toughness. The wheat varieties having very low values for dough stability and strength, in particular the soft wheat varieties were also found not so ideal for making chapattis because of their stiff or semi-stiff texture, becoming coarse and drying up rapidly after baking.

Shurpalekar and Prabhavathi (1976) successfully adapted the Brabender farinograph for objective evaluation of rheological properties of whole wheat flour dough for chapatti making. Using a 50g mixing bowl in lever position 1:3 instead of 1:1 of the farinograph, per cent water absorption and mixing time for any wheat flour could be determined by running a farinogram at the optimum dough consistency of 500 ± 20 BU. Optimum conditions for preparation of a chapatti dough were: 1 min of initial mixing, followed by a 5 min rest period, and final mixing until 1 to 2 min after the chapatti dough had attained the maximum consistency. The farinograph-mixed dough could be rolled into chapattis immediately, thereby avoiding the conventional rest period of about 1 hr. The rolling properties of chapatti dough in respect of stretching and spreading were also objectively determined using the Henry Simon Research Extensometer and the Hilliff Chapati Press, respectively. Six wheat varieties were tested for making chapattis of acceptable quality based on the rheological properties. Water absorption of chapatti dough ranged between 50 and 57 per cent, which was attributed to the varietal differences with respect to quantity and quality of gluten, and the extent of starch damage during milling because hammer mill was used to mill the flour. The dough mixing time varied between 2 min for soft wheats and 5 min for medium hard wheats. For a given variety of wheat the farinograph-mixed dough had higher extensibility and lower resistance to extension values compared to Hobart-mixed doughs. The farinograph-mixed doughs from the varieties Sonalika and Punjab (commercial) had excessive extensibility and poor resistance to extension. The spread area of rolled chapatti dough sheet in respect of farinograph-mixed dough was 20 to 30 per cent greater than that of Hobart-mixed doughs. According to these authors, the chapattis prepared from the varieties Kalayan Sona and K68 were graded excellent, and varieties Sharbati Sonora and Sonalika also yielded chapattis of good quality.

Dough Consistency: Chapatti dough consistency is an important parameter to ascertain ease of rolling and retention of circular shape of the chapatti dough sheet, and the quality of chapatti (Rao, 1982). An objective method to measure the consistency of whole wheat flour (WWF) chapatti dough using a simple Research Water Absorption Meter (RWAM) was standardized based on the extrusion principle (Rao, 1982; Rao et al, 1986a). The desired consistency of chapatti dough was measured as extrusion time, optimum being 60 ± 5 seconds when extruded with an additional weight of 2 kg after resting the dough for 20 min. The dough consistency determined by RWAM was significantly correlated to that measured by the farinograph and hardness measured in a General Foods Texturometer. The water required to prepare chapatti dough of optimum consistency was determined from a semi-logarithmic plot of the extrusion time of doughs of varying water contents against the water added. Chapatti water absorption was determined by multiplying the farinograph water measured at 500 BU by a factor of 0.85 or by adjusting the farinograph consistency to 450 ± 20 BU at a lever position of 1:3 instead of 1:1.The consistency of chapatti dough is affected by the quantity of WWF, wheat variety, amount of water, temperature, type of mixer, addition of optionally used ingredients such as common salt and or fat, etc. Chapatti water absorption, on the other hand, was influenced by flour particle size, protein and damaged starch contents. A multiple regression equation was also derived for predicting the chapatti dough water absorption based on particle size, damaged starch and the protein content of whole wheat flour. The effect of dough consistency on dough properties and chapatti quality as reported by Rao et al (1986a) is shown in **Table 9.**

Table 9: Effect of dough consistency on dough properties and chapatti quality

Water added %	Consistency as extrusion time, seconds	Dough handling and rolling	Shear value lb	Puffed height 10*	Pliability 10*	Appea-rance 20*	Hand-feel 10*	Eating quality 25*	Taste 25*	Total score 100*
60	257	Stiff, lumpy, difficult	9.2	5.3 **(4.3)**	4.0 **(1.5)**	9.4	4.4	9.5	14.0	45.6
68	97	Moderately Soft, cohesive, Easy	6.4	7.6 **(6.5)**	7.0 **(2.0)**	12.1	7.2	14.9	16.1	65.3
71	65	Soft, cohesive, Easy	5.7	8.2 **(7.3)**	8.0 **(2.1)**	14.5	8.5	18.1	16.4	73.7
72	55	Soft, cohesive, Easy	5.7	8.3 **(7.4)**	9.0 **(2.3)**	14.0	8.8	19.0	16.9	76.0
74	46	Very soft, slightly sticky, somewhat easy	5.4	8.5 **(7.6)**	9.5 **(2.4)**	14.4	9.1	21.2	16.1	76.8
80	22	Very soft, cohesive, sticky, difficult	4.2	8.5 **(7.5)**	10 **(2.5)**	12.4	9.4	23	14.0	77.3

Source: Rao et al (1986a).*Maximum score. Values in brackets indicate puffed height (cm) and pliability (cm).

The studies conducted by Jha et al (2010) on the rheological and baking properties of a diverse set of advanced Winter x Spring lines of wheat found a significant negative correlation of the chapatti water absorption with farinograph tolerance index, bread water absorption and stored chapatti moisture content, which pointed to the fact that chapattis prepared from stronger gluten wheat retrograde quickly by losing moisture at a rapid rate. On the other hand, positive correlation with fresh chapatti moisture content indicated that the extent of water absorption during dough mixing influences the extent of water retention in fresh chapatti. The higher moisture absorption in chapatti dough is desirable as it leads to a higher retention of moisture in chapattis thereby making such chapattis more soft, pliable and tastier. Moreover, the higher moisture content in chapattis is helpful in maintaining pliable texture during extended shelf life.

Effect of Supplements and Additives on the Rheological Properties

The rheological properties of wheat flour are reported to modify when supplemented with flours from other cereals or legumes, and functional additives used with the objective to improve the nutritional quality and functional or sensory properties of the end product. Following reports substantiate this:

Sidhu et al (1990a) reported that blending 25 per cent hull-less barley flour into whole wheat meal decreased water absorption, increased dough development time and dough stability but dough softening decreased. With 50 per cent barley supplemented wheat flour, the dough was smooth and extensible but showed a reduced resistance to extension.

Sharma et al (1999) showed that blending of dehulled cowpea flour to wheat flour significantly reduced water absorption, decreased dough development time, dough stability and mixing tolerance index, and increased dough softening with increasing amount of cowpea flour (**Table 11**).

Table 10: Effect of blending cowpea flour with wheat flour on rheological properties of dough

Cowpea flour (g / 100g)	Water absorption (%)	Dough development time (min)	Mixing tolerance index (BU)	Dough stability (min)	Degree of softening (BU)
0	62.5	2.3	80	1.9	70
5	62.0	2.1	75	1.9	75
10	61.5	2.1	65	1.8	80
15	61.2	1.9	65	1.7	90
20	60.5	1.85	60	1.5	90
25	60.5	1.7	60	1.5	95

Source: Sharma et al (1999).

These changes in dough properties upon addition of cowpea flour may be attributed to dilution of gluten forming proteins causing weakening of the dough, and possibly because of variation in hydration behavior of the two types of proteins. Incorporation of cowpea flour also lowered the gelatinization temperature, peak viscosity and peak viscosity temperature. The changes in pasting characteristics may have been due to reduced ability of cowpea flour granules to swell before physical breakdown and also to the reduced starch concentration in blends containing cowpea flour.

Shortenings, surfactants, hydrocolloids, enzymes, etc. used widely for improving the texture and keeping quality of baked products are also reported to influence the rheological properties of the dough. Sidhu et al (1989) used bakery shortening (partially hydrogenated peanut oil), and surfactants glycerol monostearate (GMS) and sodium steroyl-lactylate (SSL) to improve the quality of whole wheat flour chapatti. The GMS alone or in combination with shortening decreased the dough development time and the dough stability. On the other hand, SSL alone or in the presence of shortening increased the dough development time and the dough stability. The GMS also increased the dough breakdown during mixing as it gave higher drop off values. However, the drop off values were lower for shortening alone or in combination with SSL compared to the control, the GMS addition increased the peak consistency, but a decrease in peak value was observed with SSL or shortening alone and with their combination. These farinographic results were supported by the observations of chapatti dough handling characteristics. The chapatti dough prepared with surfactants alone or in combination with shortening was found to be quite smooth and extensible without being sticky, and possessed very good sheeting properties.

Addition of hydrocolloids to whole wheat flour (Ghodke and Ananthanarayan, 2007) resulted in increase in water absorption; the highest water absorption was observed for hydroxypropylmethylcellulose (HPMC) and carboxymethylcellulose (CMC) added at 1.0 per cent level. The dough development time also increased with all hydrocolloids used, Some hydrocolloids clearly influenced the dough stability; guar gum showed increase in dough stability with increasing concentration of the gum, whereas CMC showed decrease with increasing concentration.

Akhtar et al (2009) evaluated the rheological changes taking place in the dough as a result of addition of elemental iron, NaFeEDTA, zinc sulphate and zinc oxide in different combinations to whole wheat flour packaged in polypropylene woven bags and tin boxes and stored for 60 days under ambient (35-42^{0}C) and controlled (23-25^{0}C, 45-55 per cent RH) conditions. The fortificants significantly affected water absorption capacity (WAC), dough development time (DDT), dough stability (DS) and viscographic properties of flours. During storage period WAC and DDT of flours increased progressively, while storage conditions had only slightly affected the viscographic properties of the flours. Packaging materials also influenced WAC, DDT and DS. These changes in rheological properties of flours are indicative of the quality of the end use product and also help in understanding the dough handling properties during processing.

Based on the above discussion on the physicochemical and rheological properties of different wheat varieties and whole wheat flour, some of the properties that designate a wheat variety suitable for good-quality chapatti making are summarized in **Table 11.**

Table 11: Properties of whole wheat flour suitable for good-quality chapatti making

Property		Value
Grain hardness	**N**	70- 90
Flour particle size	**µ**	130- 300
Protein	%	10- 13
Ash	%	1.2- 1.5
Glutenin	%	3- 5
Gliadin	%	4- 6
Total sugar	%	2.5- 4.0
Damaged starch	%	12- 15
Diatatic activity	**mg/10g flour**	250- 375
Sedimentation value	**ml**	25- 37
Pleschenke value	**min**	120- 150
Farinograph water absorption	%	70- 80

Chapter 3

Composite Flours for Chapatti

Introduction

Cereals have dominated the dietary pattern in Asia especially the Indian subcontinent. Since cereals are deficient in several essential amino acids and micronutrients, there has been wide prevalence of protein – calorie malnutrition owing to poor bioavailability of these nutrients caused by the presence of antinutritional factors such as phytates, condensed tannins, etc. Therefore, attempts have been made to fortify cereals with other food materials that are rich in these nutrients leading to development of composite foods. The bakery products have been prepared from blends of flours derived from cereals, wheat in particular plus other food materials. The blends could be binary or ternary mixtures of flours. The term 'composite flour' was thus coined and has been defined variously by different authors:

1. As a mixture of flours, starches and other ingredients intended to replace wheat flour totally or partially in baked and pasta products (De Reuter, 1978),
2. As a combination of wheat and non-wheat flours for the production of leavened and non-leavened baked products (Dandy, 1992)
3. As a mixture of flours from tubers rich in starch (e.g. cassava, yam, sweet potato) and / or protein rich flours (e.g. soy, peanut) and / or cereals (e.g. rice, maize, millets, buckwheat) with or without wheat flour (Seibel, 2011).

The use of blends of wheat and non-wheat flours for baked products has always occurred in times of scarcity of wheat from whatever cause, climatic or economic. The main concern has been to improve the nutritional quality, and encourage the use of local non-wheat raw materials for baked products. There has been a growing interest in fortifying wheat flour with high lysine materials especially legumes to improve the amino acid balance of baked products like bread and chapatti. Other reasons include the production of alternative baking flours for people who cannot tolerate wheat products. Much of the research work aimed at incorporating non-wheat edible materials of local origin in baked products (e.g. bread, chapatti) was undertaken in India during early 1950s (Bains et al, 1952; Subrahmanyan et al, 1958). In the 1960s and 1970s, composite flours very often found themselves at the focus of attention in European and International cereal research. Most of the studies in this field were supported by the Food and Agriculture Organization of the United Nations to encourage the use of local raw materials in the developing countries. The International Association for Cereal Science and

Technology also took up this topic at an early stage (.Chatelanat, 1973). Since then, considerable interest has been generated in fortifying wheat flour with high protein-high lysine materials (e.g. legumes, oilseed flours, protein concentrates, and isolates) to increase the protein content and improve the essential amino acid balance of the baked products.

Supplementing non-wheat flours encounters technological difficulties and impairment of baking quality. The preparation of baked goods is intrinsically linked to the rheological properties of the dough. Wheat is the main ingredient in most of the baked goods. Wheat is equipped with viscoelastic gluten proteins which confer specific rheological properties to the dough, and in turn influence the final quality of the baked product. During the preparation of composite flours and doughs, partial replacement of wheat with another component that may be devoid of gluten, would influence rheological properties of the resultant dough. Generally, the major effect on dough rheological properties arises from dilution of the gluten content on partially replacing the wheat component. Various rheological parameters that are of interest particularly when the process needs to be mechanized include water absorption, dough development time, and dough strength and mixing tolerance, etc. There are numerous research studies underway on the subject of improving dough rheology of composite flours and doughs, but noticeable commercial breakthrough is still eluding. However, success may be achieved if the overall quality (odour / flavour, appearance, chewing properties, and shelf-life) of baked products made from composite flours is as similar as possible to those baked products made from wheat alone (Seibel, 2011).

Chapatti also known as *roti* in vernacular language has been the principal staple food in Indian subcontinent and has been usually made from wheat flour. However, in rural areas, *roti* is commonly prepared using cereals singly or combination of wheat flour and millets or wheat flour and maize or wheat and Bengal gram flour, etc. Chapatti or *roti* constitutes a major portion of the total food intake. The cereal proteins are generally incomplete qualitatively being limiting in essential amino acids especially lysine. Thus such heavy dependence on cereals as protein source has resulted in the prevalence of protein calorie malnutrition particularly among the poor strata of the population. Therefore, supplementing cereal flours with other vegetable or animal proteins could be one of the most promising courses of action to alleviate protein related deficiency in the subcontinent. Food materials rich in essential amino acids may be the right candidates to supplement wheat flour or other cereal flours to develop composite flours for preparing chapattis and related products. The chapattis made from such supplemented flours could readily find their way into the dietary pattern of the target population without changing their food habits. But the success of development of supplemented flours for chapatti and related products ultimately depends upon the acceptability of the product; otherwise even the most nutritious food is of no value to the malnourished. The non-wheat edible materials that have been used to prepare composite flours for chapattis include tubers rich in starch (cassava/tapioca, yam, potato, sweet potato), cereals other than wheat (rice, maize, sorghum, millets, barley, oats, triticale and pseudocereals), oilseeds (peanut/groundnut, sunflower, cottonseed), legumes / pulses (soybean, chickpea, pigeon pea, cowpea, kidney bean (*rajmah*) , cereal brans, seed husks and hulls, fruits, vegetables and their products, etc. The studies reporting the composition, rheological and chapatti-making properties, and nutritional significance of composite flours are presented.

Tuber based Composite Flours

The objective of using starch rich tubers in composite with wheat has been to offset the cost of wheat and to encourage the use of local raw materials for baked products for routine consumption. The starch rich tubers except perhaps potato are deficient in protein and other nutrients (Table 1).

Table 1: Chemical composition of flours of tapioca and potato (per 100g).

Parameter		Tapioca	Potato (dry basis)
Moisture	g	11.4	6.0
Protein (Nx6.25)g	g	1.7	9.2
Fat	g	0.3	0.4
Fibre	g	1.9	1.6
Ash	g	1.4	3.5
Carbohydrate (by diff.)	g	83.1	79.3
Vitamin C	mg	-	36.9
Nicotinic acid	mg	1.8	-
Thiamin	mg	0.12	0.04
Riboflavin	mg	0.11	-
Phosphorus	mg	80.0	172.5
Calcium	mg	60.0	28.5
Iron	mg	-	7.2

Source: Subrahmanyan et al (1958); Roy Choudhuri et al (1963).

***Use ofCassava/Tapioca*:** Subrahmanyan et al (1958) blended tapioca flour (TF) derived from cassava (*Manihot esculenta*) with whole wheat flour (WWF) in the following proportions: (a) 75 parts WWF and 25 parts TF, (b) 70 parts WWF plus 25 parts TF and 5 parts groundnut flour. Chapattis made from these blends were equally acceptable as those prepared from WWF alone. Chapattis prepared from blends containing up to 15 per cent tapioca flour were reported fairly comparable with wheat chapattis in colour, puffing, texture and sweetness (Murty and Austin, 1963). Raja (1986) used plain and parboiled cassava (tapioca) flour and blended it with whole wheat flour (WWF) in the ratio of 7:3, 1:1, and 3:7, respectively. Blending plain cassava flour to WWF at levels of 30 per cent and above improved the colour and appearance of chapattis, but scored less to handfeel, mouthfeel, and taste compared to chapattis prepared from WWF. However, blending of parboiled cassava flour with WWF improved the organoleptic qualities of the chapattis. Use of parboiled cassava flour increased the firmness and cohesiveness of the dough. In a study on wheat—cassava composite flours, Khalil et al. (2000) reported that there was a progressive reduction in water absorption of flours with an increase in substitution with cassava flour, possibly due to reduced protein content of the composite flours. Mixing time, stability, and strength of the composite dough also increased as the proportion of cassava flour increased. The addition of malt at 1% decreased mixing time and improved dough stability.Jisha et al (2008) investigated improving the nutritional and functional attributes of cassava through fortification with cereal and/ or legume flours, bran sources etc. and through pre-treatment with enzymes to improve the functionality and reduce the energy content. The study led to the development of cassava based composite flours with low starch digestibility, high protein content and low energy content which could be effectively utilized for developing designer foods for obese and diabetic people. Enhanced digestibility of pre- gelatinized malted flours from cassava finds potential application for the development of foods for geriatric and convalescent people. The composite flours thus developed may be used to prepare chapattis.

***Use of Potato*:** Potato is one of the most important and versatile foods from nutritional standpoint.Potatoes contain antioxidants, predominantly vitamin C and certain carotenoids and anthocyanins. Potato protein has favorable lysine content in comparison with cereal proteins. In terms of quantities required to maintain nitrogen balance in adult humans, the potato protein has better nutrition value than the proteins in wheat, rice or corn. The lysine contents in potatoes are similar to those in animal protein, and potato flour can be utilized to overcome protein-calorie malnutrition (Anjum et al, 2008).The use of potato in bread making can be traced back to ancient times. In ordinary white bread, 2 to 3 per cent potato flour has been used, and is reported to help retain freshness of bread and impart a distinctive flavour. Similar attempts have been made to fortify chapattis with potato (Ezekiel and Singh, 211). Mashed potato is commonly used in filled/stuffed *parathas.* Singh et al (2005) reported that incorporation of potato flour in whole wheat flour resulted in chapattis with improved water absorption/ retention, and consequently softness of chapattis. The addition of 2 per cent potato flour to wheat flour brought about a marked increase in the extensibility of chapattis and it increased progressively with increasing levels of potato flour up to 8 per cent. The addition of potato flour to wheat flour also affected the force to rupture, modulus of deformation, and energy to rupture. It was concluded that the properties of chapattis prepared from wheat flour are significantly affected by the addition of potato flour up to 8 per cent resulting in chapattis with higher extensibility, lower peak force, and lower energy to rupture. This may be attributed to lower amylose content, higher water absorption index and consistency coefficients of potato flours used. Khaliduzzaman et al (2010) prepared chapattis from the composite flour containing 80 per cent wheat flour and 20 per cent potato flour (blanched plus KMS treated). The chapattis showed the highest width, spread ratio, yield and moisture content. The higher spread ratio they considered was associated with good texture, tearability, chewiness and other quality attributes of chapatti. Chapattis made from this blend were rated high for colour, flavour, texture and overall acceptability as compared to other blends. Chemical composition of potato flour, wheat flour and supplemented chapatti is presented in **Table 2**.

Table 2: Chemical composition of potato flour (blanched), wheat flour and chapatti.

Parameter		Potato flour	Wheat flour	Chapatti (20% potato flour)
Moisture	%	10.0	13.0	34.0
Protein	%	5.0	11.5	9.4
Fat	%	0.86	1.0	0.93
Ash	%	2.50	0.70	3.70
Crude fibre	%	5.50	2.50	19.20
Carbohydrate (by diff)	%	81.64	73.80	51.97
Vitamin C (mg/100g)		5.0	-	-

Source: Khaliduzzaman et al (2010).

Other Cereal-based Composite Flours

Cereals form the staple food of majority of the people of the Indian subcontinent providing between 70 to 80 per cent of the total calories and more than 66 per cent of the protein in the diet (Juneja et al, 1980). Unfortunately, cereal proteins are deficient in several essential amino acids needed by man for proper growth. The cereals other than aestivum wheat do not make chapattis of acceptable quality. Blending cereals with one another may improve the acceptability of chapattis and the nutritional value. Since germinated wheat does not yield chapatti of acceptable quality, blending up to 20 per cent of the sound ungerminated wheat is reported to produce non-sticky dough and fully puffed chapatti during baking (Sharma et al, 1988).

***Use of Durum Wheat*:** Non-aestivum wheat such as durum wheat do not make acceptable chapattis as they puff partially, have relatively rough texture, and possess creamish yellow color (Sharma et al, 1999).. Nevertheless, in some parts of India small amount of durum wheat is mixed into the aestivum wheat to prepare chapattis of acceptable quality. Prabhavathi et al (1976) have shown that 15 to 35 per cent durum flour blends with aestivum flour yield excellent sweet chapattis, which though become stiff too early upon storage. According to Kaur and Hira (1988), the durum/aestivum chapattis made from the blend of 40:60, were rated very good for overall acceptability which was significantly higher than all other combinations and also better than aestivum and durum wheat chapattis singly. However, the chapattis made from 40:60 durum/ aestivum wheat blend supplemented with Bengal gram and soybean individually received lower overall acceptability scores and the scores further decreased with the increasing level of Bengal gram or soybean. Bakshi et al (1989) reported that chapattis made from blends of aestivum wheat and durum wheat (75:25 and 50:50) possessed acceptable creamish colour, full puffing, and were rated very good.

***Use of* Triticale**: Triticale, a hybrid cereal produced by the cross breeding of wheat and rye, is rich in lysine and also has a higher protein content as compared to wheat. But, chapattis made from triticale flour have poor acceptability due to the reddish colour and stickiness of chapatti on chewing (Anon, 1976). The overall quality of chapattis prepared from the blend of triticale plus wheat flour (40:60) was found acceptable, and addition of Bengal gram to the blend improved the texture of chapattis (Juneja et al, 1980). A blend of triticale *atta* with wheat *atta* in 1:3 ratio yielded as good a chapatti as that from wheat *atta* alone (Sekhon et al, 1980). Chapattis made from four varieties of triticale admixed with wheat in 30:70. 40: 60 and 50:50 proportions were adjudged acceptable (Chawla and Kapoor, 1982). As the proportion of triticale flours increased, the protein quality of chapattis improved. According to Bakshi et al (1989), blends (50:25:25, 33:33:33, 25:50:25) containing bread wheat, durum wheat and triticale, respectively produced chapattis with creamish colour, partial to full puffing, and overall scoring very good to good. Inclusion of triticale results in increased water absorption and decreased dough stability.

***Use of* Barley:** Sood et al (1992) reported that addition of hull-less barley flour to wheat flour increased the protein content, and raised water absorption capacity of the composite flour. Puffing of chapattis in all composite flours was satisfactory. Colour, appearance and texture of chapattis were good up to 30 per cent hull-less barley flour in the blends, but flavour score was slightly decreased. But, chewability of chapatti was acceptable up to 40 per cent of hull-less barley flour in the blend. Sidhu et al, 1990a), on the other hand, reported that hull-less barley flour from the varieties tested, can be supplemented up to 25 per cent level to wheat flour without affecting the quality of chapatti adversely. According to Anjum et al (1991), blending barley flour up to 20 per cent into the wheat flour yielded chapattis of acceptable quality. Gujral and Pathak (2002) observed that blending barley flour with wheat flour increased the extensibility of chapatti at 20 per cent level of addition but at higher levels it started to decrease. After 24 h of storage the extensibility remained significantly higher than the control sample, which may be attributed to high level of β-glucan in barley flour. Beta glucan is reported to be involved in preventing the retrogradation of starch and thus keeping the chapatti more extensible (Bhatty, 1995). Chapatti containing barley flour showed higher peak force to rupture and energy to rupture, which remained higher even after 24 h storage. Later, Gujral and Gaur (2005) prepared chapattis from the composite flours containing wheat and barley flours (0:0, 90:10 and 80:20) admixed with glycerol monostearate (0, 0.25 and 0.5 per cent w/w) and sodium chloride (0, 0.5 and 1.0 per cent w/w) and tested for texture instrumentally based on tensile deformation before and after storage. The extensibility of control chapatti decreased whereas load to rupture and deformation modulus increased upon storage. Barley flour and glycerol monostearate and sodium chloride at all levels significantly affected the instrumental texture of chapatti. All the three ingredients prevented staling of chapatti as evident from the higher extensibility values after 24 h storage. Further, Sharma and Gujral (2014b) studied the behaviour of barley flour (28, 56 and 84 per cent) or β-glucan (1.5, 3.0 and 4.5 per cent) addition to whole wheat flour on the staling of chapatti. The dough water absorption increased significantly up to 76.7 and 78.3 per cent upon incorporation of barley flour and β-glucan, respectively. Bake loss significantly increased upon incorporation of barley flour but was not significantly affected by β-glucan. The peak viscosity and final viscosity significantly increased upon incorporation of barley flour, whereas incorporation of β-glucan had the opposite effect. The pasting and thermal behavior was affected by both barley flour and β-glucan. The stored chapattis exhibited higher pasting viscosities compared to the fresh chapattis. Barley flour and β-glucan both were instrumental in delaying the staling of chapatti thus extending its shelf-life. Thondre and Henry (2009) showed that supplementation of whole wheat flour with a commercial barley β-glucan fiber preparation created palatable chapattis for diabetic patients and the glycemic index (GI) of chapattis with 4 g of β-glucan per serving was significantly reduced.

***Use of Rice*:** Gujral and Pathak (2002) substituted whole wheat flour with rice flour at 10 to 50 per cent level to prepare chapattis. Rice flour increased the extensibility of the chapatti at lower levels but at higher levels the extensibility progressively decreased because of the dilution of gluten proteins in the composite flour as rice flour lacks gluten proteins. However, increasing levels of rice flour lead to a decrease in both the peak force to rupture, energy to rupture and modulus of deformation of fresh chapatti but after 24 h storage the modulus of deformation significantly increased with increasing levels of rice flour

***Use of Millet*:** Pearl millet (*bajra*) is rich in protein, fat, iron, thiamin, and numerous bioactive compounds. According to Murty and Austin (1963), blending pearl millet to the extent of 15 per cent with whole wheat flour yielded chapattis that were creamy in colour, had rapid and full puff, possessed soft, silky and pliable texture, and were very sweet in taste. But, chapattis containing higher levels of *bajra* were grey in colour, stiff in texture, and did not puff well. However, Vijayakumar et al (2009) reported that 30 per cent millet flour incorporated composite flour-based chapatti was highly acceptable. Nevertheless, the cooking time was longer and the grainy appearance with brown colour was not appealing to the consumer. Gujral and Pathak (2002) reported that mixing millet flour in whole wheat flour at 10 per cent level significantly increased the extensibility of chapatti but at higher levels of addition the extensibility decreased. The decrease may be attributed to the dilution of gluten proteins. After 24 h storage, chapattis containing millet flour remained more extensible as compared to control sample. Millet flour brought about an increase in the peak force to rupture, modulus of deformation and energy to rupture.

***Use of sorghum*: Sorghum** (*Sorghum vulgare*) is a good source of protein, B group of vitamins like thiamine, riboflavin, B6 and niacin, minerals like iron, potassium, phosphorus and zinc, and a wide range of bioactive compounds. However, the protein quality of sorghum grain is poor because of low content of essential amino acids such as lysine, tryptophan and threonine. Also, the presence of condensed tannins in some varieties of sorghum grain decrease its nutritional value because of their binding effect (Anglani, 1998). In order to improve the nutritional status and acceptability of sorghum chapatti, attempts have been made to blend sorghum with wheat or other materials. Murty and Austin (1963) mixed *jowar* (sorghum) flour at 15 to 20 per cent levels with whole wheat flour without adversely affecting palatability and other characteristics as those made from pure wheat flour. Rehman et al (2006a) used the pearled sorghum and pearled wheat flour to prepare rotis or chapattis. The composite flour containing 15 per cent pearled sorghum flour and 85 per cent pearled wheat flour showed maximum improvement in dough development time, dough stability, tolerance index and softening of dough. Higher proportion of sorghum weakened the dough mixed in a farinograph. Baking of chapattis also improved the availability of minerals and reduced the phytate content in the chapattis. Higher level of sorghum substitution in wheat flour may affect the chapatti-making quality because sorghum is deficient in gluten forming proteins which impart elastic and cohesive properties to wheat flour dough. Seleem and Omran (2014) reported that blending sorghum flour at 5, 10 and 15 per cent level in wheat flour decreased water absorption, dough stability first increased then decreased, degree of dough softening first decreased then increased, dough elasticity and extensibility first increased then decreased as the level of sorghum flour increased. The acceptability of sorghum fortified flat bread was close to that of wheat flour flat bread. However, the freshness of sorghum fortified flat bread decreased as the storage period advanced. The mineral content, iron and zinc in particular, increased with increasing level of sorghum flour. Grain sorghum is reported to be an excellent source of a diverse array of phytochemicals that have been shown to possess multiple health benefits, including protection against cellular oxidation, inflammation, cancer, atherosclerosis, and type-II diabetes (Althwab et al, 2015).

***Use of Maize (Corn)*:** Maize contains high levels of many important vitamins and minerals including thiamine, niacin, vitamin B6 and folate, and potassium, phosphorus, zinc, calcium and iron (Watson, 1997). Maize could be composited with wheat flour to overcome stickiness and rolling difficulty of the whole maize flour dough into chapatti. Gujral and Pathak (2002) have shown that incorporation of corn (maize) flour in whole wheat flour at 10 to 50 per cent levels increased the extensibility of chapatti at 10 per cent level but it decreased with increasing level of corn flour. The peak force to rupture and energy to rupture increased initially but at higher levels of corn flour, these led to a constant decrease. This might be attributed to the absence of gluten protein in corn flour, and also to higher water absorption. Chapattis containing corn flour showed lower modulus of deformation as compared to the control sample. The chapattis containing corn flour remained more extensible and soft even after 24 h of storage probably because of higher water absorption.

***Use of Oats*:** The oats maintain a promising image as nutritious, health-promoting ingredient in food applications due to its high bioactive and functional component contents, β-glucan in particular (Duss and Nyberg, 2004). Gujral et al (2013) prepared chapattis by blending defatted and heat treated oats flour at 25 and 50 per cent levels into wheat flour. The heat treatment significantly affected the antioxidant properties of oats. Incorporating oat flours to wheat flour increased the total phenolic content but lowered the antioxidant activity, however both decreased significantly upon baking. The reducing power of the oat blended flours was higher than the wheat flour and was further increased upon baking. The metal chelating activity of flour blends further increased upon baking. After baking the total flavonoid content decreased, whereas the non-enzymatic browning index significantly increased upon baking.

***Use of Pseudo*-cereal:** Amaranth (*Amaranthus paniculatus*) grain, a pseudo-cereal contains a substantial quantity of protein and fat, and is high in lysine as compared to most cereals (Singhal and Kulkarni, 1988). It is a good source of dietary fibre, minerals such as iron, magnesium, phosphorus, copper and manganese. Composite flour containing amaranth flour and wheat flour in the ratio of 1:3 produced chapattis that scored high for appearance, taste, flavour, texture and overall acceptability (Sarojini et al, 1996 b). According to Banerji et al (2017) incorporation of amaranth flour up to 50 per cent decreased per cent water absorption and dough stability with slight increase in dough stickiness. But there was overall improvement in softness, rollability, and spreadability of dough with increasing levels of amaranth flour. Based on texturometer measurements and sensory evaluation of chapattis, addition of 40 per cent of amaranth flour was, however, recommended. The chapattis prepared from this blend had significantly higher content of ash (iron, calcium and manganese), fat, protein, and lysine with improved *in vitro* digestibility.

Dietary Fibre Supplemented Flours

The dietary fibre, also referred to as prebiotic, is composed of two types of fibre – soluble and insoluble based on their solubility. The main sources of fibre intake are fresh fruits and vegetables, whole grains, pulses and beans, and nuts. Both types of fibre are found in different proportions in fibre containing foods. Good sources of soluble fibre are oats, barley, fruits and vegetables, and pulses whereas whole grain cereals are rich sources of insoluble fibre. Significant positive health effects have been attributed to dietary fibre, including cholesterol control, modulation of glucose and insulin responses, weight management, and improved gastrointestinal functions (Duss and Nyberg, 2004). However, fibre intake is commonly lower than recommended. Consequently, the development of foods with higher fibre content should be considered desirable (The American Dietetic Association, 1993). Cereals bran is perhaps the best source of dietary fibre for supplementing fibre deficient food products. From technological standpoint, chapatti and similar single layered flat breads could be the better vehicles for delivering high dietary fibre ingredients into the diet. Several food products enriched with dietary fibre have been developed such as bread (Jinshui et al, 2002; Gomez et al, 2003), breakfast foods (Duss and Nyberg, 2004), cakes (Lee et al, 2004), biscuits (Sudha et al, 2005), cookies and chapattis (Ghufran Saeed et al, 2009; Yadav et al, 2010 & 2012).It is important from food products development point of view that high fibre foods should not only supply fibre, but also provide enhanced functional properties and be tasty, thus encouraging continued high fibre intake.

***Use of Isabgol Husk*:** Isabgol (*Psyllium mucilloid*) husk is widely used as a natural fibre in the pharmaceutical and allied industries. It is found in almost every household as it is an effective home remedy for treating common gastro-intestinal troubles besides other health promoting effects. Ahluwalia and Kaur (2001) prepared chapattis after incorporating 0.5, 1.0, 1.5, 2.0, 2.5 and 3.0 per cent *isabgol* husk in *atta* (whole wheat flour). The *isabgol* husk improved the farinograph water absorption, dough development time and stability, whereas gelatinization temperature decreased and peak viscosity increased. The puffing of chapattis was partial at all levels of substitution. The chapattis recorded maximum sensory scores for colour, taste, flavour, texture, and hand feel at 2 per cent level of substitution. These studies implied that *isabgol* husk may be added to *atta* at 2 per cent level without altering the qualities of chapatti.

***Use of Wheat Bran*:** The wheat bran mainly contains insoluble fibre. Shenoy H and Prakash (2002) prepared high fiber chapatti incorporating 5, 10 and 15 per cent wheat bran to wheat flour and evaluated for acceptability and farinograph characteristics. The bran incorporated doughs were difficult to knead and sticky with no change in rolling properties. Farinograph characteristics revealed an increase in dough development time and a decrease in mixing tolerance index as the level of incorporation of bran increased. Wheat bran incorporated products scored lesser for the sensory quality attributes than the controls. The differences were, however, statistically not significant for products prepared with 5 per cent level of bran. It was concluded that wheat bran can be incorporated up to a level of 5 per cent for the preparation of high fiber chapattis or *phulkas* without affecting sensory quality. Yadav et al (2010) recommended incorporating 5.5 g bran per 100 g whole wheat flour for making optimally acceptable fibre rich chapattis. The optimized chapattis contained more of total fibre, insoluble fibre, and soluble fibre compared to the control, thus meeting FDA (1998) standards for claiming the functional health benefits (**Tables 3 & 4**). The rheological properties of composite flour dough were greatly affected with the addition of bran. The bran had negative effect on peak viscosity, break down and final viscosity, and peak mixing value. The bran had positive effect on hardness of chapatti, while the cohesiveness, chewiness and overall acceptability decreased with increase in bran content. The adverse effects of wheat bran on the rheological properties of dough and sensory quality of chapatti may be reduced through pretreatments such as particle size reduction, presoaking in water, hydrothermal treatment, use of enzymes, surfactants, etc. (Hemdane et al, 2016).

Table 3: Chemical composition (per cent) of whole wheat flour, wheat bran and oats bran

Parameter	Whole wheat flour	Wheat bran	Oats bran
Moisture	10 .6	2.2	2.6
Protein	13.4	16.5	16.8
Fat	1.3	4.9	8.6
Ash	1.6	4.3	1.5
Carbohydrate	69.3	40.7	55.0
Total dietary fibre	3.8	31.5	15.6
Insoluble dietary fibre	2.9	26.8	6.4
Soluble dietary fibre	0.9	4.7	9.2

Source: Yadav et al (2010).

Table 4: Chemical composition (per cent) of wheat bran fortified and control chapatti

Parameter	Control	Bran Fortified
Moisture	30.2	31.0
Protein	10.5	10.6
Fat	4.5	4.5
Ash	1.1	1.3
Carbohydrate	51.2	47.9
Total dietary fibre	2.7	4.7
Insoluble dietary fibre	2.1	3.4
Soluble dietary fibre	0.66	1.4

Source: Yadav et al (2010)

Dar et al (2014) prepared composite flours containing 5 and 10 per cent wheat bran which yielded fully puffed chapattis with overall acceptability score similar to that of whole wheat flour chapattis. However, the texture of bran fortified chapattis was slightly harder than that of the control as the cutting force increased with increase in substitution. The bran enriched chapattis showed non-sticky behavior during dough handling. Overall, chapattis prepared by the addition of 10 per cent bran, showed better performance and were quite comparable with whole wheat flour chapattis in respect of proximate composition and sensory attributes.

***Use of Oats Bran*:** The dietary fibre in oats bran contains more soluble fibre than any other grain. The main constituent of soluble fibre is β-glucans. The β-glucan in oats bran is highly bioactive and is equipped with several functional properties. It has the property to slow digestion, and resulting in the feeling of satiety and suppression of appetite. It is also known for reducing the absorption of glucose in the digestive system and for its hypocholestrolemic effect (Duss and Nyberg, 2004). It would be ideal to incorporate oats bran into a wide variety of food products including baked products.

Yadav et al (2010) recommended addition of 9.7 g bran per 100 g whole wheat flour for making optimally acceptable fibre rich chapattis. The optimized chapattis contained more of total fibre, insoluble fibre, and soluble fibre compared to the control, thus meeting FDA (1998) standards for claiming the functional health benefits (Table 3). The rheological properties of composite flour dough were greatly affected with the addition of bran. The bran had a positive effect on set back and final viscosity, but peak mixing value decreased with increase in the level of bran. The bran had positive effect on hardness of chapatti, while the cohesiveness, chewiness and overall acceptability decreased with increase in bran content. Dar et al (2014) also supplemented chapattis with 5 and 10 per cent oats bran yielding fully puffed chapattis with overall acceptability score similar or better than that of whole wheat flour. The bran enriched chapattis showed non-sticky behavior during dough handling. Overall, chapattis prepared by the addition of 10 per cent bran, showed better performance and were quite comparable with whole wheat flour chapattis in respect of proximate composition and sensory attributes.

***Use of Rice Bran*:** Rice bran is the outermost brown layer surrounding the endosperm of whole-grain brown rice. Rice bran contains oil and is rich in vitamins (thiamin, riboflavin, niacin, pyridoxine, pantothenic acid, biotin, and mixed tocopherols), minerals (magnesium, potassium, iron, manganese, and zinc), and a host of antioxidants (e.g., phytosterol ã-oryzanol and tocotrienols). These antioxidants in whole rice bran are particularly effective in promoting cardiovascular health, and reducing hypercholesterolemia in humans. The rice bran protein has high nutritional value with a protein efficiency ratio of 1.6. The bran contains total dietary fiber in the range of 21 to 27 per cent with less than two per cent soluble fibre. Rice bran has many food applications in prepared foods, nutraceuticals and functional foods. Some of the common applications include bakery products, snack foods, nutrition bars, cereals, pasta products, and medical foods (Hoffpauer, 2005; Kahlon, 2009). Cereals, wheat in particular, constitute major portion of the total calories in the daily diet in third world countries. Rice bran could be a good candidate for supplementing in wheat flour to enhance the nutritional value and reduce the cost of wheat based food products. In India, in spite of being abundantly available and nutritionally rich, it is under-utilized for human consumption, and is mostly used as animal/poultry feed. Ghufran Saeed et al (2009) studied the effect of addition of 2, 5, 10, 15 and 20 per cent rice bran to commercial white wheat flour on the rheological behavior of the dough using Farinograph, Consistograph and Alveograph. The physicochemical properties of the dough were not affected significantly even with the addition of 20 per cent rice bran. Chapattis prepared from wheat flour substituted with up to 10 per cent rice bran were rated acceptable in respect of colour, appearance, texture, taste and mouth feel. However, rice bran was found to cause perceptible bitter taste and dark colour in the chapattis. Yadav et al (2012) blended defatted rice bran (DRB) of different particle size with wheat flour in the proportion of 0, 5, 10, 15, 20 and 25 per cent and evaluated for dough and chapatti making quality. The extensibility of dough and chapatti decreased while peak load to rupture of chapatti increased at 25 per cent replacement of wheat flour with fine DRB. The water absorption capacity of wheat flour increased with 25 per cent DRB, the highest increase being in the case of fine DRB substitution. The overall acceptability score of chapatti also decreased significantly with 20 per cent DRB, the decrease being lowest for the fine DRB blend. The chapattis prepared from the wheat flour containing 20 per cent fine DRB were rated acceptable and also had significantly higher ash and total dietary fiber content than the control chapattis. Singh et al (2013) blended the microwave stabilized rice bran with whole wheat flour in the ratio of 20:80 to prepare chapattis. The rice bran fortified chapattis were equally acceptable as those prepared from whole wheat flour. The glycemic index of rice bran based chapattis was significantly lower than the controlled chapattis. Dar et al (2014) reported that composite flours containing 5 per cent rice bran yielded fully puffed chapattis but puffing was partial with 10 per cent bran. However, the overall acceptability score was close to that of whole wheat flour chapattis. The texture of bran (both levels) fortified chapattis was slightly harder than that of the control as the cutting force increased with increase in substitution. The bran enriched chapattis showed sticky behavior during dough handling. Overall, chapattis prepared by the addition of 5 per cent bran showed better performance and were quite comparable with whole wheat flour chapattis as for as sensory attributes were concerned. Free fatty acids content of rice bran fortified chapattis was more than the control and it increased with increasing level of rice bran.

***Use of Soy Bran / Fiber / Hulls*:** Soy bran / fiber are produced by toasting and grinding seed coat portion of the soybean. This bran has a crude fiber content of approximately 38 per cent, one of the highest fiber sources among commercially available vegetable products. It finds its applications in multigrain breads and other bakery products, and in several blended fiber systems (Dubois and Hoover, 1981). Khan et al (2005a) blended Soy hulls (bran) milled flour into wheat flour at 1.5, 3, 4.5, 6 and 7.5 per cent levels. Supplementation significantly increased minerals (calcium, magnesium, manganese, iron and zinc) content which was also reflected in chapattis prepared from the respective flours. Phytic acid content was found to decrease in chapattis during baking. The chapattis prepared from the composite flour containing 4.5 per cent soy hulls flour were found acceptable to the panel of judges in respect of appearance, colour, flavour, taste, texture, chewability and foldability. The fiber content of chapattis also increased with increase in substitution.

***Use of Fiber Mixture*:** Arya and Sonawane (2016) reported that addition of a mixture of wheat and soy fibers, and type III resistant starch (RS) in wheat flour decreased dough stickiness and improved dough strength. Reduction in dough stickiness may prove beneficial to food industry in terms of savings in energy. The dough containing mixture of 2.5, 5.0 and 2.5 per cent wheat fibre, soy fibre and type III RS, respectively yielded highly acceptable chapattis in terms of textural properties such as low tear force values and high extensibility.

Legume / Pulse based Composite Fours

Pulses and lentils are consumed as accompaniment to the cereal based staple diet of the majority of the population of the Indian subcontinent. Pulses are the richest source of plant proteins and provide approximately 10 per cent of the total dietary requirements of the protein world over. Pulses are also high in minerals, dietary fibers and complex carbohydrates leading to low glycemic index foods. Pulses help to lower cholesterol and triglycerides as leguminous fibers are hypoglycosuria because of consisting of more amylose than amylopectin. One of the most important properties of legume protein is their high content of essential amino acids, e.g. lysine, and their deficiency in sulphur containing amino acids making them a potential complement to cereals, e.g. wheat. Pulses, therefore, provide tremendous opportunities to be utilized in the processed and prepared foods (Asif et al, 2013). The use of composite flours containing legume flours at suitable levels in baked products has been projected (Marnett et al, 1978). However, supplementation with non-wheat flours involves technological difficulties and impairment of baking quality. Unit operations employed to produce legume/pulse flours impact both positively and negatively on the nutritional, physicochemical and functional properties of the flours produced. The presence of anti-nutritional factors such as trypsin and chymotrypsin inhibitors, oligosaccharides, lectins and tannins also limit the availability of nutrients present in legumes. Nevertheless, soaking, germination, dehulling and cooking are reported to improve the nutritional and functional properties of legumes (Ghavidel and Prakash, 2006). This eventually influences the sensory quality of foods in which legume-cereal composite flours are used. The ratio of flours in the composite may influence dough rheology and consequently physical and sensory characteristics of baked products thereof. This directly influences consumer acceptability of final products. It, therefore, becomes pertinent to design the entire process of producing composite flours critically so that the overall quality (sensory as well as shelf-life) of baked products made from composite flours is as similar as possible to those baked products made from wheat alone (Seibel, 2011).

***Use of* Soybean:** Soybean with 40 per cent proteins and 20 per cent oil though has a great potential in solving the problems of protein-calorie malnutrition, it has not been the legume of choice because of its beany flavour and trypsin inhibitor activity. Nevertheless, the beany flavour and the trypsin inhibitor activity can be reduced or eliminated by boiling or autoclaving the soybean prior to use in composite flours. Soybean is rich in lysine, which is a limiting amino acid in most cereals. Although soybean protein is low in sulphur containing amino acids, supplementation with cereals which are well supplied with sulphur containing amino acids will result in a composite product in which the quality of protein will be high and well balanced (Rathod and Williams, 1970). Soy flour in bakery products is not only used for nutritional reasons but also for its functional properties. Shyamala and Kennedy (1962) prepared chapattis from a mixture of 90 per cent whole wheat flour and 10 per cent defatted soybean flour. The soybean fortified chapattis did not differ appreciably from those made with all whole-wheat flour in colour, texture and flavour. There was a marked improvement in the protein quality of the chapattis resulting in protein efficiency ratio over 50 per cent greater than for the whole wheat chapattis. Khan et al (1972) used different protein-rich flours to fortify whole wheat flour admixed with 2 per cent common salt and 0.5 percent sodium steroyl-lactilate for the preparation of chapattis. Soy fortified chapattis were softer as tested with precision Penetrometer. The organoleptic tests showed that chapattis made from 5 percent soy-blend received higher scores compared to 10 and 15 per cent blends. Rathod and Williams (1973) investigated the consumer acceptance of wheat-soybean blends to ascertain the levels of full fat, partial fat, and defatted soybean flour to prepare acceptable chapattis. Defatted soybean-wheat flour (20:80) combination was preferred for chapatti making. Chapattis prepared by Bhat and Vivian (1980) using whole wheat flour admixed with 10 and 20 per cent soybean flour received high scores in respect of colour, appearance, texture, flavour and overall acceptability, and were nutritionally superior in respect of total protein and lysine content. Gandhi et al (1983) demonstrated that chapattis prepared from wheat flour containing 10 to 15 per cent soybean flour were of satisfactory quality in terms of flavour, texture, appearance and overall acceptability, but flavor and texture were significantly different from the all-wheat flour chapattis. Incorporation of defatted soybean flour in wheat flour has been reported to have softening effect on chapattis and balances the amino acid pattern of chapatti without affecting its acceptability characteristics (Ebeler and Walker, 1983; Lindell and Walker, 1984). Verma et al (1987) prepared full fat soybean flour after steaming beans for 5 min. at 15 psi to reduce trypsin inhibitor activity to minimum while maintaining maximum nitrogen solubility index. The results of sensory evaluation of chapattis prepared from different wheat-soybean blends indicated that chapattis containing 25 per cent full fat soybean flour were found neither acceptable nor disliked, but those with 10 per cent full fat soybean flour were quite acceptable. The effect of added soybean paste and instant soy *dhal* to wheat flour on softening of fresh chapattis, and substantial reduction in toughness which normally develops during storage was studied using Instron Universal Testing Machine (Gandhi and Bourne, 1988). Instant soy *dhal* and soy paste, each added to wheat flour at levels of 10 and 20 per cent reduced the puncture force by 25 and 33 percent, and 23 and 38 percent, respectively. Upon storage for 24 h the puncture force of all-wheat chapattis increased by 318 per cent, being very tough and unacceptable. In contrast, after 24 h the 10 and 20 per cent soy *dhal* products had, respectively mean puncture forces of 58 and 37 per cent of the 24 h all-wheat chapattis. Those prepared with soybean paste were even more resistant to ageing; the 10 and 20 per cent soybean paste products had puncture force only 39 and 30 per cent of the all-wheat chapattis, respectively, and the latter is still softer than the freshly prepared all-wheat chapattis. Chauhan and Santosh Kumari (1990) reported that the chapattis made from the composite dough containing 10 per cent defatted soy flour were comparable to control in sensory attributes and showed complete inactivation of trypsin inhibitor activity with slight degradation of phytate and 4 to 5 folds increase in reducing and non-reducing sugars. The protein content of chapattis increased from 12.6 to 20 per cent, and moisture retention from 29.6 to 37.7

per cent indicating maintenance of freshness of chapattis for a longer period. Rawat et al (1994) studied the effect of soy-fortification on the physicochemical, nutritional and textural properties of chapattis. Blending 10 per cent defatted soybean flour with whole wheat flour yielded chapattis with soft texture owing to greater moisture retention compared to whole wheat flour chapattis. Soy- fortification enhanced the protein and mineral content of chapattis and stepped up the PER, but only marginally improved the digestibility. However, fortified chapattis retained 13 per cent of the trypsin inhibitor activity originally present in defatted soybean flour (Table 5).

Table 5: Chemical composition of whole wheat flour (WWF), defatted soy flour (DSF) and Chapattis.

Parameter / Product		WWF	WWF chapatti	DSF	20 % DSF- chapatti
Moisture	%	-	31.5	-	33.5
Protein	%	12.4	11.8	57.1	15.2
Fat	%	1.5	1.8	0.7	1.4
Ash	%	1.0	1.7	7.4	2.5
Crude fiber	%	0.7	1.5	3.4	2.4
Carbohydrates (by difference)	%	84.4	83.2	31.4	78.5
Reducing sugars	%	0.6	0.5	1.2	0,5
Non-reducing sugars	%	1.7	1.7	3.3	1.9
Calcium	mg/100g	120.0	106.7	311.4	209.1
Phosphorus	mg/100g	150.0	194.2	698.6	283.1
Iron	mg/100g	2.6	3.4	16.0	4.7
Available lysine,	g/16g N	2.2	2.1	6.0	2.5
Trypsin inhibitor activity	TUI/mg	-	-	0.3	0.04

Source: Rawat et al (1994).

Bajwa (1997) studied the rheological properties of dough and textural characteristics of chapattis as affected by 5 to 20 per cent replacement of wheat flour with defatted soy meal, and storage for 4 and 24 h at 25° C. Addition of defatted soy meal increased water absorption, dough development time and mixing tolerance index, and decreased dough stability. The Instron shear-force increased and deformation to peak shear-force decreased significantly with increased addition of soy meal. The textural characteristics of chapattis were also affected significantly upon storage. Incorporation of 10 per cent soymeal was found optimum in terms of increased protein quantity and quality without any appreciable effect on dough properties and chapatti characteristics. Gandhi et al (2000) reported that mixing defatted soy flour at 20 per cent level in whole wheat flour had significant effect on the nutritional and textural properties of chapattis. The protein content increased from 11.9 to 19.8 per cent. Chapattis were acceptable when fresh but not after 24 h. The textural properties like hardness, cohesiveness, springiness and chewiness of chapattis were also affected by the addition of defatted soy flour both at zero and 24 h storage. Deshpande et al (2001) first dehulled and treated the soybeans to detoxify and remove the beany flavor before milling into full fat soy flour. Soy-blended chapattis were prepared from the composite flour containing wheat flour and full fat soy flour in the ratio of 90: 10, respectively. Chapattis received high scores for taste, flavour, colour, texture, appearance and overall acceptability by a panel of trained judges. Khan et al (2005b) prepared composite flours by replacing wheat flour with 8, 16, 24, 32 and 40 per cent soy flour. Chapattis were evaluated for sensory quality (colour, flavour, taste, texture, foldability and chewability) to assess the acceptance of suitable level of substitution. Chapattis were acceptable by the panel of judges up to 24 per cent substitution. There was a significant increase in minerals (iron, zinc, copper, manganese, calcium and magnesium) and phytic acid content of the composite flour. However, there was a significant decrease in phytate during baking of chapattis. Khetarpaul and Goyal (2008) prepared chapattis using partially defatted and salt treated soy *dhal*, wheat and rice flour. Sensory evaluation for different parameters, i.e. colour, appearance, taste, flavour, texture and overall acceptability showed that chapattis were acceptable. The protein and total sugar contents of chapattis were significantly higher than those of the control. The processing treatments, i.e. mixing and roasting involved in chapatti making reduced the phytic acid and polyphenol contents significantly over the unprocessed composite flour. As a result, in vitro protein and starch digestibility of chapattis also improved significantly. Kadam et al (2012) treated the soybean grains by first soaking in water for 4-6 h and then autoclaving at 5 psi. The sun dried grains were pulverized into fine flour. The full fat soy flour was blended with wheat flour in different proportions. The chapattis made from composite flour containing 10 per cent soybean flour recorded high sensory quality scores. However, at higher concentration the values of sensory scores were found to decrease. The protein content and quality, and mineral content increased with the addition of soybean flour. Supplementation of this blend with dried fenugreek leaves powder at 5 percent level further increased the nutritional quality of chapattis particularly in minerals (calcium and iron) and fibre. Bhise and Kaur (2013) used texturized defatted soybean meal at 10, 20, 30 and 40 percent levels in blend with wheat flour (*atta*) for making chapattis. Chapattis were evaluated for physical, textural, sensory and functional properties. The cutting force of chapattis increased with increased level of texturized soymeal. Colour, flavour, taste, texture and overall acceptability of chapattis were negatively affected when level of texturized flour increased as compared with the control. The colour of chapattis darkened with increased level of substitution. The overall acceptability scores were maximum for control and chapattis with 10 per cent texturized meal. The functional properties such as water absorption index, foaming capacity and protein digestibility increased while water solubility index and fat absorption capacity decreased with increased level of soymeal.

Use of Chickpea (*Cicer arietinum*)**:** Bengal gram or chickpea flour is commonly used in blend with whole wheat flour by many households in the Indian Subcontinent to make chapatti known as *missi roti*. Murty and Austin (1963) used chickpea flour in blend with wheat flour at various levels to prepare chapattis. The chapattis acquired yellowish colour. Even though the flavor of gram was very prominent, a blend containing up to 15 or 20 per cent Bengal gram flour was acceptable for making chapattis. Chapattis puffed rapidly and full, were soft and pliable and sweetish in taste. Singh et al (1991) studied the improvement of functional and baking properties of wheat-chickpea flour blends with addition of gluten at different levels and sodium steroyl-lactilate (SSL) at 0.5 per cent level. Chickpea flour at 10 and 20 per cent level improved the dough stability and degree of softening which were adversely affected at 30 per cent level of blending. Puffing ability and softness of chapattis deteriorated with increase in chickpea flour beyond 20 per cent level, but improved by the addition of SSL Sharma et al (1995) found that the chickpea flour supplemented flat bread had lower moisture and harder texture as compared with the unsupplemented one; the hardness increased further during storage. Sensory evaluation revealed that superior quality flat bread could be produced up to 20 per cent chickpea flour supplementation. The protein efficiency ratio of supplemented flat bread was significantly higher than that of the control.

Siddique et al (1996) used 10 per cent undecorticated chickpea flour admixed with 90 per cent whole wheat flour to prepare chapattis which were rated highly acceptable. There was significant improvement in biological value, net protein utilization, and net protein retention and protein efficiency ratio. Lysine, calcium, magnesium, iron, copper and manganese content also improved. The rheological properties, however, were not affected by the addition of chickpea flour. Gujral and Pathak (2002) incorporated 0, 5, 10, 15, 20 and 30 per cent Bengal gram flour in to whole wheat flour for making chapattis. Water absorption increased with the increasing level of Bengal gram flour. Substitution of Bengal gram flour at 5.0 per cent level increased the extensibility of chapatti but at higher levels the extensibility was lower than that of the control. The increasing levels of Bengal gram flour decreased the peak force to rupture but increased the energy to rupture and the modulus of deformation. The chapatti containing Bengal gram flour remained more extensible and soft than the control even after 24 h of storage.

Shahzadi et al, (2005) used composite flour containing commercial wheat flour (*resultant atta*) plus chickpea flour (0, 5, 7.5 and 10 per cent) which exhibited increase in water absorption, dough development time and dough stability with increasing level of chickpea flour. However, during the period of 60 days storage of composite flours water absorption decreased and dough development time increased. Guar gum addition in the composite flour influenced the rheological properties. The sensory attributes of chapattis in terms of colour, flavour, taste, texture, chewability and foldability decreased as the storage period advanced, but were acceptable. Addition of guar gum (1 per cent) was found to impart puffiness and a whiter look to chapattis.

Gill et al (2005) investigated the effect of chickpea, *ghee*, sodium chloride, mixing time and resting time on instrumental texture and sensory quality of whole wheat flour *chapatti*. The wheat-chickpea composite flours contained 0, 5, 10, 15 and 20 percent decorticated chickpea flour, 4, 5, 6, and 7 per cent *ghee* (clarified butter), and 0.5, 1.0, 1.5 and 2.0 per cent sodium chloride. The chapattis prepared were tested for extensibility, peak force to rupture, modulus of deformation and energy to rupture using Instron Universal Testing Machine after 0 and 24 h of storage. The chapattis were also examined for puffability, appearance, colour, mouthfeel, flavour and overall acceptability. Fresh chapattis were soft and extensible as indicated by low force values required to deform, low modulus of deformation and longer distances of extension before rupture. After 24 h of storage, chapattis became stale, brittle and hard as shown by high force values, high modulus of deformation and short distances of extension before rupture. The composite flour blend containing 15, 7 and 1.5 percent chickpea flour, *ghee* and sodium chloride, respectively with a mixing and resting times of 6 and 40 min, respectively was found most suitable for preparing chapattis possessing highest extensibility as well as sensory scores. The independent variables employed in the investigation had significant effect on textural and sensory parameters of chapattis.

Kaur (2009) reported developing a Ready Mix Formulation containing wheat *atta*, Bengal gram flour and several functional ingredients for *missi roti.* The optimum level of wheat *atta* was incorporated with Bengal gram 50, fenugreek powder 5, fat 2, salt 1.5 , red chili powder 0.25, glycerol monostearate and sodium steroyl-lactilate 0.25 per cent, respectively. Pathania et al (2017) later modified the condiments formulation including onion flakes, green chilli powder and mango powder to enhance the taste and flavour. Two formulations were prepared, one with fat and the other without fat. The composite flours were evaluated for chapatti making quality and sensory characteristics. Both the formulations were packed in different packaging materials and stored at ambient and refrigeration temperature and tested for changes in the proximate composition, water activity, free fatty acids, peroxide value and texture of chapattis for a period of four months. The aluminum laminate packaging was found more suitable compared to zip pouches and high density polyethylene. Storage under refrigerated temperature was found best for both the formulations.

Kadam et al (2012) used pearled chickpea flour to blend with wheat flour at levels of 5, 10, 15, 20, 25 and 30 per cent for the preparation of chapattis. Sensory evaluation showed that chapattis made with 20 per cent chickpea flour received the highest scores for colour and appearance, flavour, taste, texture and overall acceptability. Blends containing 5, 10 and 15 percent chickpea flour also yielded equally good chapattis. However, chapattis made from blends containing higher than 20 per cent chickpea flour received lower scores for texture and overall acceptability.

***Use of Cowpea*:** Cowpea (*Vigna sinensis*) is creamy white soft textured bean with a black eye and is known as *lobia* in vernacular language. It contains around 25 per cent protein and is rich in amino acid, lysine. It is an excellent source of fiber and is rich in potassium, vitamins A and C. It has a good amount of other minerals (calcium, magnesium, phosphorus, iron, zinc, copper, manganese and selenium) and vitamins (thiamine, riboflavin, niacin, vitamin B6, folate and pantothenic acid). The high lysine content (486 mg / g nitrogen) makes cowpea an excellent enhancer of protein quality when combined with cereal grain proteins which are low in lysine but rich in sulfur amino acids (Prinyawiwatkul et al, 1996). Despite its high protein content and amino acid lysine, the cowpea as a food source has not been utilized especially in developed countries to its full potential. The constraint has been the presence of anti-nutritional factors, e.g. trypsin inhibitor, and also the presence of flatulence causing oligosaccharides particularly raffinose and stachyose in cowpeas (Sosulski et al, 1982). The beany flavor in cowpea has been reported to decrease by soaking in acidified water, dehulling, and blanching in100^{0}C steam (Okaka et al, 1979). Sharma et al (1999) prepared chapattis from decorticated cowpea flour in blend with wheat flour in the ratios 0: 100, 5: 95, 10: 90, 15: 85, 20: 80 and 25: 75. The farinograph studies showed that blending of cowpea flour in wheat flour decreased water absorption significantly. However, Mustafa et al (1986) reported an increase in water absorption of flour on blending with cowpea meal containing hulls in it. The reason may be attributed to the absence of hulls in the cowpea flour used in the present studies. Dough development time, dough stability and mixing tolerance index decreased whereas dough softening increased significantly with increasing level of cowpea flour in the blends. Dough handling was not affected up to 10 per cent substitution, but beyond this level of substitution, the dough became sticky and difficult to sheet, and puffing of chapatti was partial. The changes in dough characteristics upon addition of cowpea flour may be attributed to dilution of gluten-forming proteins causing weakening of dough. Sticky dough indicates that the water absorption rate is low. A longer holding time (after mixing) before sheeting may overcome the problem. The overall acceptability score of chapattis with the addition of 5 per cent cowpea flour was higher than that of the control but with further substitution the overall acceptability scores progressively decreased (**Table 6**). The colour and pliability of chapattis was affected adversely with increasing levels of cowpea flour in the blends. Substitution beyond 10 per cent cowpea flour resulted in chapattis with a yellowish colour and hard and brittle texture. However, these kinds of adverse effects may be overcome by the addition of suitable surfactants (Singh et al, 1991). Incorporation of cowpea flour increased the protein, ash, calcium and phosphorus contents of the blends considerably.

Table 6: Effect of blending cowpea flour in whole wheat flour on chapatti-making properties

Cowpea flour (%)	Water absorption (%)	Dough handling	Puffing	Acceptability Score (out of 9.0)
0	69.5	Nonsticky	Full	7.98
5	69.0	Nonsticky	Full	8.15
10	68.0	Nonsticky	Full	7.43
15	68.0	Slightly sticky	Full	7.20
20	66.5	Sticky	Partial	6.60
25	66.0	Sticky	Partial	5.80

Source: Sharma et al (1999).

***Use of Rajmah* (Kidney bean):** Rajmah (*Phaseolus vulgaris*) is traditionally consumed as whole legume in Northern parts of India. Composite flour was prepared using dehulled, split and milled *rajmah* in blend with wheat flour in the ratio of 25:75. Dough making quality of composite flour was similar to wheat flour, and dough was soft and elastic and could be rolled easily. The textural qualities of chapattis as judged by chewing and eating were similar to wheat chapatti. The chapattis remained soft even after 5 h of preparation. The sensory scores for appearance, flavor, taste, texture, and overall acceptability of the composite flour chapatti were almost the same as assigned to wheat flour chapatti. The composite flour could be stored safely for 12 weeks, and chapattis made from stored composite flour were rated higher than the chapattis made from fresh composite flour. The reason could be reduction in the beany flavor of the composite flour upon storage (Sarojini et al, 1996 a).

***Use of Ricebean*:** Ricebean (*Vigna umbellate*) is rich in protein, available lysine, calcium and zinc. Kaur and Mehta (1992) blended ricebean flour with wheat flour at 10, 20 and 30 per cent levels to prepare chapattis. Blending ricebean flour up to 10 per cent level with wheat flour improved all the organoleptic characteristics (e.g. appearance, colour, taste and flavour, texture, doneness, overall acceptability). However, as the level of ricebean flour in the blend increased, chapattis were scored lower for organoleptic characteristics. Results also revealed increase in protein, ash, crude fibre, minerals (calcium, iron zinc, and copper) content. There was a significant increase in available lysine, a limiting amino acid in wheat.

***Use of Pigeonpea*:** Pigeonpea (*Cajanus cajan*) supplemented wheat flour at 10 and 20 percent level showed decreased water absorption capacity; however dough development time and stability increased with increased level of pigeonpea flour. Dough handling was satisfactory at 10 per cent supplementation but the dough became sticky with addition of more of pigeonpea flour. The supplemented flat bread had lower moisture and harder texture as compared with the control; the hardness increased further during storage. Sensory evaluation revealed that superior quality flat bread could be produced up to 10 per cent pigeonpea flour supplementation. The protein efficiency ratio of supplemented flat bread was significantly higher than that of the control (Sharma et al, 1995). Harinder et al (1999) used dehulled pigeonpea flour which is rich in protein and iron (**Table 7**). Whole wheat meal was blended with 5, 10, 15, 20 and 25 per cent pigeopea flour. Increased level of pigeonpea flour decreased water absorption of whole wheat meal. At higher levels of pigeonpea flour supplementation, dough handling was made difficult due to stickiness. Chapatti puffing was full up to 15 per cent substitution; any further increase caused partial puffing. Increased level of substitution imparted a yellow colour and hard texture to the chapattis. Chapattis prepared from 5 to 15 per cent supplemented flours were acceptable. The nutrient content particularly protein, iron and phosphorus increased with increased substitution.

Table 7: Chemical composition (per 100g) of pigeon pea and whole wheat meal

Parameter/Product		Pigeon pea	Whole wheat meal
Protein	g	25.90	9.40
Fat	g	1.96	1.54
Ash	g	5.52	1.35
Total sugar (as glucose)	g	4.12	5.19
Phosphorus	mg	248.80	228.70
Magnesium	mg	68.00	75.00
Iron	mg	8.20	3.41
Copper	mg	1.08	0.40
Sodium	mg	1.24	1.17
Potassium	mg	8.20	5.33

Source: Harinder et al (1999).

***Use of Khesari*:** Khesari (*Lathyrus sativus* L) known as Indian vetch, is valued as a nutritious legume primarily owing to its high protein content (26.5 to 28.7 per cent) and especially as a good source of some of the essential amino acids (Kay, 1979). Although, it is tasty to eat besides being nutritious, it has a drawback because of the presence of a neurotoxin. However, it can be detoxified prior to use by steeping in hot water and repeated rinsing. Once detoxified, it can be used as *dhal* (split grain); its flour may be blended with wheat or other cereal flours for making bakery products. Rehman et al (2006b) prepared composite flours by mixing resultant *atta* (roller mill flour) with detoxified *khesari* flour in 0, 10, 20, and 30 per cent ratio. The protein content of composite flours increased with increasing level of *khesari* flour. Chapattis prepared from different blends were evaluated for sensory quality and biological value. The sensory attributes like color, taste, flavor, texture, breakability and foldability revealed that the chapattis were most acceptable to taste panel when supplemented with up to 20 per cent *khesari* flour. The biological value, net protein utilization, digestibility, net protein efficiency ratio increased and feed gain ratio decreased as a result of improvement in protein quality and depletion of the toxin.

***Use of Fenugreek Seeds*:** Fenugreek (*Trigonella foenum-graecum* L.) leaves (both green and dried) and seeds form an important ingredient in Indian cooking. The green *methi* leaves *paratha* is particularly famous for its delicious taste and aroma. The seeds are generally used as condiment and have medicinal properties especially being hypocholesterolemic and hypoglycemic. The seeds though bitter in taste, can be debittered to increase their acceptability (Sharma, 1984). Fenugreek seeds are rich in proteins, available lysine, and mucilaginous and total fibre. As fenugreek seeds are rich in mucilaginous fibre and other dietary essentials, their use can be exploited as functional and nutritional food as well as therapeutic agent. Sharma and Chauhan (2000) studied the effect of debittered fenugreek seed flour on the physicochemical and rheological quality characteristics of fenugreek supplemented wheat flour, and acceptability of bread thereof. Darak Priya and Pawar (2015) prepared chapattis with partial replacement of wheat flour with fenugreek malt (seeds germinated and dried) at 5, 10, 15 and 20 per cent levels. Chapattis were evaluated for physicochemical, sensory, nutritional and microbial characteristics. There was significant increase in protein and ash content and decrease in carbohydrate content. The blend containing 15 per cent fenugreek malt received the highest sensory scores. The colour of chapattis was brownish yellow as measured with colorimeter and Munsell colour chart. The texture of chapatti was measured on a texture analyzer. Increasing levels of fenugreek malt lowered the peak force to rupture and increased the modulus of deformation. The mucilaginous fibres in fenugreek may be responsible for the softening effect as these have higher capacity to retain moisture. Calcium propionate used to control mold growth during storage, was found to improve the overall acceptability score of 15 per cent malt fortified chapattis. The fenugreek malt was also reported to reduce total serum cholesterol, total lipids and LDL cholesterol.

Oilseed Based Composite Flours

The use of protein rich full fat or defatted oilseed meals/flours shows promise in improving the nutritive value of the supplemented products, at the same time optimum utilization of the meals /flours.Oilseed flours/meals are reported to significantly improve the nutritional quality of wheat flour because of their high contents of protein, essential fatty acids and essential amino acids especially lysine (Bhat and Vivian, 1980). Several attempts have been made to enrich chapatti using oilseed flours without sacrificing its functional and sensory characteristics.

***Use of Peanut Flour*:** Bhat and Vivian (1980) prepared chapattis from whole wheat flour with 0, 10 and 20 per cent substitution with defatted peanut flour. The chapattis incorporated with peanut flours received similar sensory scores for appearance, colour, texture, flavour, and overall acceptability as those for control. The scores were well above the minimum acceptable score of five for all the characteristics and overall acceptability. The protein content and available lysine contents of chapattis increased with increased level of peanut flour.

***Use of Cottonseed Flour*:** Bass and Caul (1972) prepared composite blends replacing wheat flour with 5, 10 and 15 per cent cottonseed flour (CSF). All the supplemented flours and doughs were darker than the control and contained black specks as well as the aroma contributed by CSF. The doughs were initially dry and subsequently became inelastic. The chapattis prepared from these blends were darker than the control and possessed CSF aroma. Gumminess, partial-puffing and suppressed wheaty aroma character of chapattis from the blends may also limit the use of CSF.Bhat and Vivian (1980) prepared chapattis from whole wheat flour with 0, 10 and 20 per cent substitution with defatted cottonseed flour. The dough with 10 per cent cottonseed flour possessed a greenish tinge and was darker than the plain wheat flour dough. The intensity of greenish colour and darkness of dough increased with increased level of cottonseed flour. The chapattis containing cottonseed flour received significantly lower sensory scores for appearance, colour, flavour, texture and overall acceptability than those for control. The chapattis became dry and leathery within a very short time. Incorporation of cottonseed flour, however, increased the protein and available lysine contents of chapattis. These results suggest that the cottonseed flour used needs to be further refined to eliminate the undesirable colour and flavour for preparation of acceptable chapattis.

Use of Flaxseed: Flaxseed or linseed (*Linum usitatissimum*) is a pleasant tasting, slightly nutty and sweet whole grain. Flaxseed is rich in nutraceutical components including dietary fibre, lignin, and polyunsaturated fatty acids especially α-linolenic acid (omega-3 fatty acid) an essential fatty acid, which human body cannot manufacture from any other substance. Flaxseed has several health benefits including prevention of cardiovascular disease, stroke, cancer, diabetes, osteoporosis, improvement of immune function, and laxative properties (Wu et al, 2007). It can be readily incorporated into a wide variety of foods especially cereal based food products including baked goods, breakfast cereals, pasta, energy bars, dry mixes for pancakes, and even beverages.

Preethi B and Chimmad (2014) prepared omega-3 enriched chapattis from the composite flour obtained by grinding together wheat and 8 per cent linseeds. The enriched chapattis puffed partially and had irregular baking spots; however, the texture was soft. The chapattis were evaluated for colour, appearance, texture, taste, flavor and overall acceptability. Except for creamish brown colour, the other sensory parameters were at par with the control. The dough of the enriched flour was slightly sticky which may be due to the presence of mucilage in linseed. The dough was less elastic compared to the control possibly due to the disruption of the gluten network by linseed particles. Incorporation of linseed resulted in increase in protein, fat, dietary fiber and fatty acids content especially omega-3 fatty acid compared to the control. The enriched chapatti flour was storable up to two months at ambient temperature, but the enriched chapattis received lower sensory scores as storage period advanced from zero to 24 hr. Bhise and Kaur (2013) used texturized defatted flaxseed meal at 10, 20, 30 and 40 percent levels in blend with wheat flour (*atta*) for making chapattis. Chapattis fortified with texturized flaxseed meal showed increased contents of protein, fat, fiber and ash, and rise in protein digestibility as compared to the control. Chapattis were evaluated for physical, textural, sensory and functional properties. The cutting force which reflects the texture of chapattis, increased with increased level of texturized flaxseed meal. Colour, flavour, taste, texture and overall acceptability of chapattis were negatively affected when level of texturized flour increased as compared with the control. The colour of chapattis darkened with increased level of substitution. The overall acceptability scores were maximum for control and chapattis with 10 per cent texturized meal. The functional properties such as water absorption index, foaming capacity increased while water solubility index and fat absorption capacity decreased with increased level of flaxseed meal. However, chapattis were acceptable for all attributes evaluated up to 10 per cent substitution while higher levels had negative effect. The studies demonstrated that deoiled cake from flaxseed oil industry has potential for supplementation of protein and other valuable nutrients in food products. Manjula Sasikumar (2015) prepared *roti* from composite flours containing 20. 40, 60, and 80 per cent flaxseed powder in blend with wheat flour. The dough rolling ability and separation of layers during puffing were examined. The organoleptic properties of *roti* were tested in terms of appearance, colour, texture, flavour, taste, softness and overall acceptability. The *roti* prepared from the blend containing 60 per cent flaxseed flour received the highest scores for overall acceptability.

***Use of Sunflower Seed Meal*:** Sunflower seed meal is a palatable source of high quality protein and fiber. The kernels are an excellent source of vitamin E and phytosterols, very good source of selenium, vitamin B1, and a good source of magnesium, potassium, manganese, copper, phosphorus, vitamin B6, folate, niacin, phytochemicals, as well as essential fatty acids (Holiday and Phillips, 2001). Dehulled seeds are relished in the toasted form, and are used in bakery and confectionary products, and in gluten free recipes. Bhise and Kaur (2013) used texturized defatted sunflower seed meal at 10, 20, 30 and 40 percent levels in blend with wheat flour (*atta*) for making chapattis. Chapattis fortified with texturized sunflower seed meal showed increased contents of protein, fat, fiber and ash, and rise in protein digestibility as compared to the control. Chapattis were evaluated for physical, textural, sensory and functional properties. The cutting force of chapattis increased with increased level of texturized sunflower seed meal. Cutting force reflects the texture of chapatti. Colour, flavour, taste, texture and overall acceptability of chapattis were negatively affected when level of texturized meal increased as compared with the control. The colour of chapattis as determined using a Minolta Spectrophotometer darkened with increased level of substitution. The overall acceptability scores were maximum for control and chapattis with 10 per cent texturized meal. The functional properties such as water absorption index, foaming capacity increased while water solubility index and fat absorption capacity decreased with increased level of sunflower seed meal. However, chapattis were acceptable for all attributes evaluated up to 10 per cent substitution while higher levels had negative effect. The studies demonstrated that deoiled cake from sunflower seed oil industry has potential for supplementation of protein and other valuable nutrients in food products.

Fruits and Their Products based Composite Flours

***Use of Jamun*:** *Jamun* (*Syzygium cumini*, L) also known as Indian blackberry is rich in minerals, vitamins, phytochemicals like flavonoids and anthocyanins, hence designated as neutraceutical. It has several health benefits and possesses antioxidant, antidiabetic and anticancer properties (Chaudhary and Mukhopadhyay, 2012). Kapoor et al (2015) incorporated conventional hot-air-dried (CD) and freeze-dried (FD) *jamun* powder at 1, 2, 3, 4, and 5 per cent levels into whole wheat flour for preparing supplemented chapattis. Water absorption was found to increase with increased addition of *jamun* powder. Chapatti dough was non-sticky with 1 to 4 and 1 to 3 per cent levels of CD and FD *jamun* powder, respectively. Supplemented chapattis puffed fully at all levels of substitution except at 5 per cent that resulting in partial puffing. The colour and texture of chapattis were determined using Hunter Lab Colorimeter and Texture Analyzer, respectively. FD *jamun* powder supplemented chapattis were softer compared to those supplemented with CD *jamun* powder; may be due to high hygroscopicity and higher absorption capacity of FD *jamun* powder. Chapatti colour darkened with progression in supplementation levels. There were differences in sensory properties of CD and FD *jamun* powder supplemented chapattis. Flavour was found to be a prominent factor in determining acceptability of supplemented chapattis, and astringency was found to increase with increased substitution. Among different substitution levels, 3 per cent level of substitution received maximum overall acceptability scores for both CD and FD *jamun* powder supplemented chapattis. The bioactive components like ascorbic acid, total anthocyanins and total phenols were retained better in FD *jamun* powder.

***Use of Dates*:** Waghray et al (2011) prepared chapattis from the composite flours containing whole wheat flour and dates paste in the ratio of 70:30, 60:40 and 50:50. The chapattis were evaluated for colour appearance, texture, taste, flavour, mouth feel, after taste, and overall acceptability. The composite flour chapattis admixed with 30 per cent dates received higher scores for all the sensory attributes compared to that of control. However, increased addition of dates progressively lowered the overall acceptability of the supplemented chapattis.

***Use of Jackfruit Seeds*:** Jackfruit (*Artocarpus heterophyllus* L.) seeds are rich in protein, starch, minerals, B-group of vitamins and phenolic compounds. Tulyathan et al (2002) have shown that jack fruit seed flour has good water and oil absorption capacities. Starch isolated from the flour has high amylase content and good paste stability. Substitution of wheat flour with the seed flour reduced the gluten strength of the mixed dough. Traditionally, seeds are consumed as snack after boiling or roasting. The ground flour from dried or roasted seeds is blended with wheat flour for baking. This gluten-free seed flour can serve as a substitute for wheat flour for those with specific food allergies. Ocloo et al (2010) report the composition of jackfruit seed flour (per cent dry matter basis) as protein13.50, fat 1.27, ash 2.70, fibre 3.19, carbohydrate 79.34, and calcium 308mg, iron 13 mg, potassium 1478 mg, sodium 6 mg, copper I mg with a calorific value 382.79 kcal/100g. The flour possesses high water and fat absorption capacity. The studies conducted by Chowdhury et al (2012) have shown that jackfruit seeds flour in blend with wheat flour can be used for developing bakery products and as protein supplement and functional ingredient in food formulations. The flour being rich in starch may also be used as thickening and binding agent in food systems. Sultana et al (2014) prepared chapattis from composite flours made of whole wheat flour + white jackfruit seed (lye-peeled) flour + brown jackfruit seed flour + Bengal gram flour in the proportions of 80:5:10:5, 55:15;15;15 and 45:25:20:10 and the control which were fortified with same amount of fat, salt, sugar and baking powder. The colour of chapattis changed from light brown (control) to dark brown by the addition of jackfruit seed flour. The texture of chapattis softened with the addition of jackfruit seed flour compared to the control. The spread ratio (width over thickness) of chapattis increased with increased addition of jackfruit seed flour. The protein, ash, crude fibre and fat content of chapattis increased with increased level of jackfruit seed flour. The chapattis exhibited a storage life of 3 to 4 days at ambient temperature and 30 days at refrigerated temperature. It was concluded that chapatti containing 80 per cent whole wheat flour, 5 per cent white jackfruit seed flour, 10 per cent brown jackfruit seed flour plus 5 per cent Bengal gram flour was the best among the three blends because of its higher calorific value and overall sensory acceptability (see **Table 8** for composition).

Table 8: Composition (per cent) of whole wheat flour (WWF), jackfruit seed flour (JSF), Bengal gram flour (BGF) and JSF fortified chapatti (JFC).

Parameter	WWF	JSF	BGF	JFC
Moisture	13.0	8.10	8.09	20.70
Protein	10.5	11.43	24.50	11.63
Fat	0.9	1.98	6.10	1.28
Ash	1.3	3.92	3.10	1.72
Crude fibre	1.7	2.87	1.30	1.79
Carbohydrates (by difference)	74.3	74.57	58.21	64.67

Source: Compiled from Sultana et al (2014).

***Use of Banana Peel*:** Fruit peels are good sources of dietary fibre, polyphenols, carotenoids and other bioactive compounds which possess various beneficial effects on human health. Banana peel is rich in dietary fibre, proteins, minerals especially calcium, potassium and iron, and possesses antioxidant properties owing to the presence of polyphenolic compounds (anthocyanins, flavonoids, etc.) and carotenoids (Nagarajaiah and Prakash, 2011). Kurhade et al (2016) prepared chapattis from the composite flours incorporating 0, 5, 10, 15 and 20 per cent banana peel powder (BPP) into whole wheat flour and examined for bioactive compounds, microstructural quality and sensory acceptability. The proximate composition of BPP showed high levels of dietary fibre, ash and protein (**Table 9**). With increased level of BPP, dough stickiness and dough strength increased. However, kneading and rolling of dough eased with increased level of BPP perhaps owing to increased water absorption. The colour of chapatti as tested using Hunter Lab Colorimeter darkened with the increased level of BPP, whereas the texture as tested on Texture Analyzer softened as indicated by the decreasing tear force. However, extensibility of chapatti declined as the level of BPP increased. The sensory evaluation showed alteration in colour and taste of chapatti with increased level of BPP. Scanning electron microscopy showed changes in starch properties due to the addition of BPP. The phytochemical content and antioxidant activity increased with the increased level of BPP in chapatti. The results reveal that the chapatti prepared from dough incorporated with BPP would be a good carrier to impart health benefits as it possesses antioxidant activity.

Table 9: Proximate composition of Banana Peel Powder (per cent dry weight basis)

Parameter	Value
Moisture	1.77
Protein	7.58
Cellulose	7.07
Ash	11.10
Fat	2.09
Total carbohydrate	77.40
Total Dietary fibre	53.80

Source: Kurhade et al (2016).

Vegetables and Their Products based Composite Flours

Green leafy vegetables are micronutrient dense nature's gift to mankind. They are a good source of high quality protein, vitamins, and minerals especially calcium and iron. They also contain an immense variety of bioactive, non-nutritive health enhancing factors such as antioxidants, phytochemicals, essential fatty acids and dietary fiber.The high supplementary value of green leafy vegetables, e.g. beet, knol-khol, turnip, *bathu* (*Chinopodium album*), carrot, radish, mustard, Chinese cabbage green leaves to wheat chapatti has been reported in Chapter on Nutritive Value (Sawhney and Kawatra, 1986; Luthra and Sadana, 1995).

***Use of Spinach*:** Khan et al (2013) prepared spinach chapattis using whole wheat flour admixed with fresh blanched seasoned and spiced spinach paste, hydrogenated fat, salt, sugar, and sorbic acid as antimycotic agent. The composition of spinach chapattis showed the presence of higher level of protein, fat and minerals especially iron. The chapattis were subjected to in-pack sterilization and stored at ambient temperature in laminated packaging material for 12 months. The colour and texture of chapatti as determined using Hunter Lab Colorimeter and Texture Analyzer respectively changed negatively during storage. There was gradual loss of carotenoids, chlorophyll and vitamin C during storage. The sensory scores for colour, aroma, taste, texture, and overall acceptability changed slightly during storage. However, it was concluded that highly acceptable and nutritious spinach chapattis could be prepared by incorporating fresh spinach paste, and their shelf-life could be extended up to 12 months by using sorbic acid and in-pack sterilization. Khan et al (2015) also developed chapatti premixes by blending whole wheat flour with 1 to 10 per cent spinach powder prepared by blanching and dehydrating spinach leaves. Addition of spinach powder resulted in increase in moisture absorption, dough peak time and tenacity, whereas peak viscosity, breakdown viscosity decreased with increasing level of spinach powder from 1 to 10 per cent. Incorporation of 5 per cent spinach powder to wheat flour was found to be optimum for chapatti making. Any further increase in the concentration of spinach powder adversely affected the texture, colour and overall acceptability of chapattis. With increasing spinach powder levels in chapattis, the Hunter colour values and textural properties in terms of hardness, chewiness, springiness and gumminess were affected. Addition of spinach powder (1 to 10 per cent) significantly increased total carotenoids, total chlorophyll, vitamin C, iron, calcium and magnesium contents in the premixes and chapattis. However, there was significant decrease in total carotenoids, total chlorophyll as well as vitamin C contents during baking of chapattis. It was concluded that highly acceptable quality chapattis may be prepared using 5 per cent dehydrated spinach powder. Although the level of dehydrated spinach powder incorporation is small, the increase in the nutrient content is remarkable. Dehydrated spinach powder supplemented chapattis, thus have the potential to serve the valuable source of calcium, iron, fibre, vitamin C and carotenoids in the diet of the consumers.

Chapter 4

Chapter Preparation and Mechanization

The process of preparation and mechanization of chapattis involves three essential steps, viz.

1. conversion of flour into the dough,
2. rolling/sheeting of the dough, and
3. baking-cum-puffing of the shaped dough sheet/disc into chapatti.

Ingredients

Whole-wheat flour (WWF), commonly known as *atta* of 85 to 100 % extraction, and water are the main ingredients used in the preparation of chapattis. Resultant *atta*, which is similar to WWF, may also be used.The resultant *atta* is a mixture of flour and considerable amount of finely powdered bran particles and germ obtained as a by-product after the separation of straight-run flour and bran during the roller milling process (Tara et al, 1969). The resultant *atta* is much coarser with higher bulk density and has somewhat uniform particle size than *atta*. It is considered nutritionally superior to *atta* for its high thiamin and riboflavin contents. The *resultant atta* is largely used by *dhaabas* (roadside eateries), restaurants, hotels, industrial / institutional canteens, etc. as it is more economical, and chapatti and *tandoori roti* made from it are more brightly looking and attractive. However, it is rarely used at household level because the chapattis dry out faster and become stiff to chew. The physicochemical characteristics of *resultant atta* and *atta*, and quality of chapatti made from them are shown in **Table 1**.

Table 1: Physicochemical characteristics of *atta* and *resultant atta*, and quality of chapatti

Characteristic / Quality		*Atta**	*Resultant Atta**
Over tailings of 10 **XX** (130μ) sieve	%	31.3	71.1
Moisture	%	8.2	10.2
Ash	%	1.55	0.84
Ether extract	%	2.10	1.47
Protein (N x 5.7)	%	9.66	9.05
Free fatty acids	%	123.2	173.6
Damaged starch	%	14.1	6.6
Chapatti water absorption	%	69.5	60.0
CHAPATTI			
Colour value	%	40.0	50.0
Wheaty aroma		Good	Mild
Chewing quality		Normal	Leathery
Taste		Sweet	Bland

Source: Rao et al (1983). *Same wheat grist used

Common salt and vegetable oil or hydrogenated vegetable oil (fat) or *ghee* (clarified butter) form the optional ingredients in chapattis and *tandoori rotis*, while they are essential in the preparation of *parathas* and stuffed *parathas*. Salt is traditionally used in households for taste as well as for improving handling and rolling properties of the dough. The use of 1.5% salt was found optimum yielding dough relatively softer to hand feel and having better rolling properties (Shurpalekar and Prabhavathi, 1976). Rao *et al* (1986a) have further shown that salt has stiffening action on dough with improved cohesiveness and rolling properties but without any stickiness. The addition of fat mellows the dough improving thereby the consistency, handling and rolling properties. The addition of optimum levels of both salt and fat, therefore, improve the overall quality of chapatti.

Traditional Method for Chapatti Preparation

In the traditional method for making chapatti, the quantity of water required for the preparation of chapatti dough, dough sheeting with respect to shape, thickness and size, and baking and puffing conditions for the dough sheet are determined by trial and error. Ease in sheeting / rolling the dough disc (raw chapatti) is normally used as a criterion for assessing the consistency of the dough. Considerable experience is required to roll and bake chapatti perfectly. The experienced cook or house wife manipulates the consistency of the dough irrespective of the quality of whole wheat flour by the judicious use of water and additives like salt and or fat. This was demonstrated by Choudhry (1968), who asked three Indian and Pakistani housewives to prepare dough from 300 g of *atta* of 95 percent extraction from US wheat. The three dough preparations were immediately tested in a farinograph. The amounts of water used by the three housewives were 49, 59 and 71.5 percent, and dough consistency was recorded as 1000, 780, and 410 BU, respectively. The housewife who used the least amount of water also used some vegetable oil in preparing the dough. Though the three dough preparations varied in consistency to such a large degree, but still could be made into acceptable chapattis.

Preparation of Dough: The practice of making dough has as many variants as there are chapatti consuming households in chapatti consuming areas of the world. Whole- wheat flour may be of coarse or fine granulation, and it may or may not be sieved to remove the coarser portion of the flour. Generally, 4 to 6 % of coarse bran is sieved off using a 30 mesh sieve (Rao et al, 1986b). The quantity of salt and fat, if used, may vary from household to household. The quantity and quality of water also varies depending on the desired stiffness of the resulting dough. Water used may be warm or at ambient temperature; generally of latter type but preferentially warm in winter. Salt and fat when used should be dispersed well in the flour before the addition of water. The kneading of dough is accomplished manually using clinched fists in a bowl or shallow, rimmed stainless steel/ brass plate; size depending on the quantity of flour to be kneaded. The period of kneading also varies depending on quantity of flour, warm or ambient temperature water used, and use of salt and or fat. Essentially, kneading is continued with the addition of water bit by bit till soft pliable dough of desired stiffness is obtained. Converting flour into dough of desired stiffness is an art and is mastered after sufficient experience. The manual kneading of dough involves pressing, folding and stretching motions facilitating incorporation of air into the dough. Well-kneaded dough is springy or elastic when pressed with a finger. Simple dough kneading machines, both manually and electrically operated, are now available to knead dough. The dough is subsequently allowed to rest at room temperature to equilibrate for adequate hydration of starch and protein. It may or may not be covered with a moist piece of cloth to prevent surface drying. Alternatively, surface of the dough is moistened and covered with a lid. The period of dough resting may vary from nil to 30 minutes or more depending on convenience. Most local chapatti makers and households find the rest period necessary for the preparation of chapatti of good quality. Unused dough may be stored in refrigerator or at ambient temperature in winter. However, in large canteens serving industrial workers or devotees in temples and *gurudwaras*, either a planetary vertical mixer or a horizontal mixer similar to the one used for bread dough, is now used. The dough is mixed for 10 to 15 minutes for complete development of the gluten, rested for a period ranging from 15 to 20 minutes. Resting of dough ensures adequate hydration and stress relaxation, which enable easy sheeting / rolling without much stickiness (Rao, 1982).

Dough Rolling / Sheeting: The dough is then knocked back by kneading. Alternatively, a suitable portion of the dough is manipulated in the palms of hands by repeated pressing, folding and rounding leading to a spherical ball locally known as '*paera*'. The weight of the dough ball varies from region to region and household to household ranging from 40g to 100g or even more depending on the size of chapatti / *roti* desired. The ball is then dusted with dry flour and shaped into a thin, flat, circular disc either by slapping between the palms of hands or by using a board and rolling pin commonly called *chakla and belan*, respectively. Dusting of the *chakla* (board) and the dough disc being rolled with flour is frequently done to prevent sticking. The thickness and diameter of the rolled dough disc may vary depending upon the individual household practice. A thickness of 2 to 3 mm and a diameter of 15 to 20 cm are common. Notwithstanding, in some parts of the Indian subcontinent (e.g. Pakistan) slightly thicker and large diameter chapattis / *rotis* are more common, particularly in villages. Sheeting of dough using a rolling pin is critical and time consuming, and needs experienced persons to make chapattis having circular shape as well as uniform thickness and regular size. Non-uniformity in thickness results in uneven baking and reduced puffing owing to crust developing cracks.

The *chakla* is a flat usually circular board, commonly made of stone or wood, size varying from 20 cm to 30 cm diameter and 2.5 cm to 5 cm thickness with weight sufficient to provide stability and reduce slippage during rolling. The *belan* (Indian style rolling pin) typically made of wood is round, roller type, cylindrical 5 to 10 cm in diameter and 20 to 30 cm in length with thinner handles which extend through the roller at both ends. Rubenthaler and Faridi (1982) have designed a laboratory moulder along with a roller pin to minimize problems associated with uniform manual flattening of flat breads.

Simple, manually or electrically operated, chapatti presses are now available to shape chapatti dough into circular disc of uniform size and thickness (IS, 1979; Kannur et al, 1985).

Baking and Puffing: The circular dough disc is then transferred on to a preheated flat or concave shaped iron griddle commonly called *tawa* which is continuously heated using fire wood / dried cow dung cake / glowing coal, or kerosene / LPG flame or electric hot plate. The choice of heating medium depends upon the economic status of the household and availability of the facility. The rolled circular dough sheet is baked on each side taking a total time of 1 to 2 minutes and then transferred on to the live fire/ flame to puff assuming a convex shape caused by steam accumulation between the bottom and top layers of the baking dough disc. Puffing may also be achieved while on *tawa* by gentle pressing and rotating circularly the baked disc using a small cloth pad. Any break or weak point in the baking dough disc surface will cause escape of steam and poor puffing. It is desirable to have uniform and even thickness with continuous layer of dough in the rolled dough disc, which largely depends upon the strength of the gluten network to hold the wet flour in continuous skin. Gluten strength is not the only criteria for good puffing as chapattis prepared from durum and dicoccum wheats though poor in gluten strength but puff to the extent of bread wheat chapatti (Misra, 1998). There are other factors that affect full puffing of chapatti. Optimum dough consistency, appropriate temperature and optimum time of baking both sides contribute greatly to full and rapid puffing of the chapatti. Over baking of the first side of the sheeted circular shaped dough disc on hot t*awa* makes it to lose more moisture resulting in less steam generation during the puffing act. Therefore, the first side of the dough disc should be turned over the moment the appearance of the top side changes from creamish to wheatish due to partial dehydration, and tiny bubbles erupt on the surface, consequently resulting in more moisture available for steam generation during puffing action. Thus baked product is called chapatti or *roti.* The thinner chapatti with a thickness of 1 to 1.5 mm, a diameter of 10 to 12 cm, and fully puffed is fondly called *phulka* (meaning fully puffed). Small quantity of edible refined oil or hydrogenated vegetable oil or *ghee* (clarified butter) or butter is applied on the first baked side of the chapatti. The chapatti is generally served fresh otherwise it becomes somewhat hard and dry losing its taste and mouth feel within a short period (30 to 60 minutes). It may be, after application of fat, stored wrapped in a piece of cloth and placed in a closed container to slow down cooling and moisture loss. The flow sheet for the traditional method of chapatti preparation is shown in **Figure 1.**

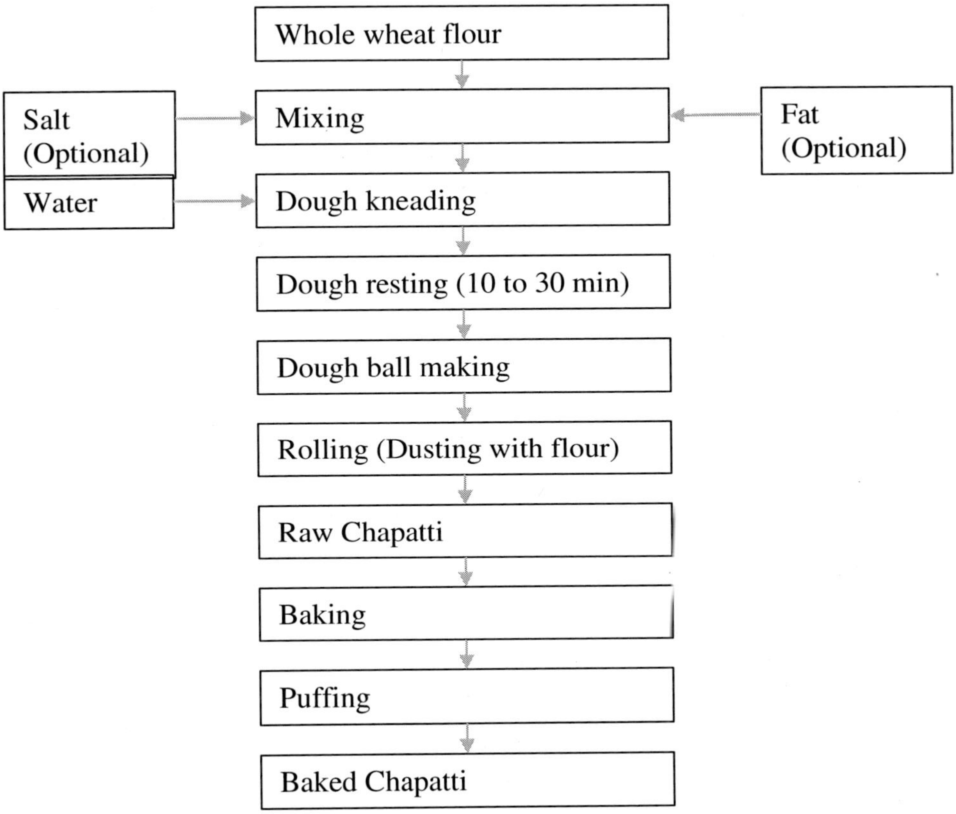

Figure 1: Traditional method flow sheet for the preparation of chapatti

Standardization of Chapatti- making Method

The conditions for the preparation of chapatti with respect to grinding wheat into whole wheat flour, addition of salt and or fat, mixing with water and dough consistency, thickness, size and shape of the dough sheet, and baking conditions (temperature and time) vary widely in different regions and laboratories (Singh and Bailey 1940, Aziz and Bhatti 1960, Choudhry 1968, Austin and Ram 1971, Shurpalekar and Parbhavathi 1976, Rao 1982, Rao et al 1986b, Navnidhi et al, 2007). Consequently, the quality of chapatti varies considerably. This situation introduces anomalies in assessing the chapatti-making quality of a given wheat variety, and evaluating the effect of other ingredients on the acceptability, nutritional value, and storage life of chapatti. Moreover, the development of objective methods of standardization and evaluation of chapatti dough and chapatti quality will be of great practical use in partly or fully mechanizing the chapatti-making process. It, therefore, became imperative to develop a standard method for test baking of chapatti, like for bread, to obtain reliable and comparable results, particularly in research and development institutions and other wheat quality control laboratories, and in mechanizing chapatti production. Some of the chapatti making standardization procedures developed at different laboratories are described below:

1. **Chaudhri and Muller** (1970) in their studies developed a standard test baking method for chapattis. Wholemeal flour (300 g) was mixed with water in a Brabender Farinograph at 25^{0}C for four minutes to a consistency of 600 BU. The dough was covered with a Perspex dish and allowed to rest at room temperature for one hour. One hundred gram portions of the dough were rounded off between the palms of the hands to balls commonly called *paeras* in vernacular language.. Each *paera* was then rolled into six inches diameter disc using a rolling pin on a board on which a six inches diameter circle had been drawn. During the process of rolling 1.5 g of dusting flour was used per *paera* to overcome sticking. The disc was then baked on an electric hotplate regulated to a temperature of 213^{0}C for four minutes until brown. After one minute of baking the chapatti was turned over and baked for one minute on the other side. It was then turned over every thirty seconds to complete four minutes of baking. The average baking temperature was recorded as 106^{0}C.
2. **Austin and Ram** (1971) proposed a standard method to make chapatti to evaluate the chapatti making quality of wheat varieties under development through breeding program at Indian Agricultural Research Institute, New Delhi. The dough was prepared by mixing 50 g whole-wheat flour (40 mesh sieve) with 30 ml water in an automatic mixer for 2 minutes. If the dough did not develop fully after 1.5 minutes, a few more ml of water was added with a pipette so that dough of moderately stiff consistency was formed during the next half a minute of mixing in the automatic dough mixer. The total amount of water added was recorded. The amount of water required to produce dough of right consistency could also be obtained more accurately on the basis of water absorption (per cent) in a farinograph. The dough thus obtained was covered with a piece of wet cloth and allowed to rest for 30 minutes in a humid chamber maintained at 30°C. The dough was then divided into two equal portions. Each portion was allowed to run for 10 times between the rotating rolls of an automatic sheeting roll and flattened into a thin uniform sheet of 3 mm in thickness. This was cut into a circular disc of 15 cm in diameter by means of a metallic circular die. The circular dough disc was weighed and placed on a hot iron plate kept in a rotary hearth baking oven which was regulated to maintain a temperature of 232.2°C. After 90 seconds, the dough disc was turned over to bake the other side for another 60 seconds. The chapatti was taken out and weighed again. Puffing usually starts after a minute of baking in the oven. Austin and Ram (1971) used this method to evaluate the chapatti making quality of newly developed wheat varieties.
3. **Bass and Caul** (1972) also developed a standard procedure to make chapattis from composite flours. Dough was prepared using flour (300g) and demineralised water (202 ml) by mixing in a Hobart N–50 mixer, initially for 1

minute at low speed and then for 3 minutes at medium speed. After mixing, the dough was placed in a Pyrex container, sprinkled with 5 ml demineralised water, covered and allowed to rest at ambient temperature for I hr. After rest period, the dough was mixed again at medium speed for additional I min and rested for 15 min. The dough was then divided into 50g pieces, which were hand-shaped into spherical balls. Each ball was flattened with a rolling pin into a disc 6.5 inches in diameter, using 1g of flour for dusting to prevent sticking. The dough disc was quickly transferred to a cast iron griddle preheated to 450^{0}F over automatically controlled gas burners. After 30 sec the disc was turned over and pressed with a soft cloth for uniform distribution of steam and separation of two layers. After another 30 sec, it was turned over again, and removed from the griddle after 30 sec taking thus a total baking time of 1.5 minutes; whenever the temperature of the griddle increased to 500^{0}F, the baking time was accordingly decreased by as much as 25 sec.

4. The above test baking methods for chapatti lacked objectivity and depended on the experience of the investigator. **Shurpalekar and Prabhavathi** (1976) demonstrated the successful use of the Brabender Farinograph, Henry Simon Research Extensometer, and Hilliff Chapati Press for standardization and objective evaluation of chapatti dough. With a 50 g mixing bowl in lever position 1:3 of the farinograph, water absorption (per cent) and mixing time for any whole-wheat flour could be determined by running a farinogram at the optimum dough consistency of 500 ± 20 BU. The optimum conditions for the preparation of chapatti- dough using 50 g flour containing 1.5per cent (optimum) common salt dissolved in water (50 - 57 per cent absorption) were found to be: 1 minute of initial mixing followed by a 5-minute rest period, and final mixing until 1 to 2 minute after the chapatti dough had attained the maximum consistency. The rolling / sheeting characteristics of chapatti dough, as reflected by its stretching and spreading properties, were objectively evaluated using the extensometer and the chapatti press, respectively. The farinograph-mixed dough could be rolled into chapattis immediately thereby avoiding the conventional rest period. The doughs prepared in Hobart mixer required 1.5 to 3.0 per cent extra water to obtain the same consistency and rolling characteristics as those of farinograph-mixed doughs. This implies that the farinograph may be used to determine the water requirement for the preparation of standard chapatti dough in different types of mixers.

5. Rao (1982) used a research water absorption meter (Henry Simon) to determine water absorption of the chapatti dough of optimum consistency and found that the amount of water required for dough to extrude in 60 to 62 seconds at a pressure of 4.4 kg weight is the measure of water absorption.

   In the standard chapatti test baking method, **Rao et al** (1986b) obtained the whole-wheat flour in a Kamas mill or other laboratory hammer mill using a 0.8 mm sieve. The flour contained 8.1 per cent moisture, 10.6 per cent protein, and 1.45 per cent ash. The water requirement for making chapatti dough of optimum consistency was determined on either a farinograph or a research water absorption meter. On the farinograph a 50g mixing bowl was used and the quantity of water required to bring the farinograph consistency of 450 BU was determined at a lever position of 1:3. The water absorption of chapatti dough using the water absorption meter was determined as described by Rao (1982). The dough of optimum consistency was obtained in a Hobart mixer (beater arm set at a low speed of 58 rpm) by mixing a weighed quantity of whole-wheat flour (14 per cent moisture) with predetermined quantity of water (67.8 to 76.6 per cent water absorption) for 3 minutes. A small piece of dough (~ 40 g) was rolled to a thickness of 2.00 mm on a rectangular platform fitted with an open rectangular aluminum frame. Frames of varied heights may be used depending on thickness of chapatti desired. However, the optimum thickness of dough sheet was found to be 2.0 to 2.5 mm to get better quality chapatti with respect to texture and overall quality. The chapatti made from 1.5 mm sheeted dough, on the other hand, had tough texture and poor overall quality. The dough was rolled with a wooden rolling pin, the length of which exceeded the width of the aluminum frame to assist application of roller force to the dough when required. The rolled dough sheet was then cut into a circular disc 15 cm diameter using a sharp-edged punching die. The circular dough sheet disc was baked on a thermostatically controlled hot plate at 205°C. After baking on one side for 45 seconds, it was turned over and baked for another 105 seconds. The baked chapatti was allowed to puff by placing in a gas *tandoor* oven maintained at 340 ± 30°C with lid closed. The optimum puffing time was 25 seconds. Immediately, the height of the puffed chapatti was measured in centimeters. The chapatti was then cooled on a hollow wooden stand for about 15 minutes and preserved in a closed container for quality evaluation. Baking chapatti on *tawa* or hot plate–cum–gas fire resulted in better texture, aroma, taste and overall quality as compared to chapatti baked on hot plate and puffed, either on live- coal fire or in gas *tandoor* (Rao, 1982).The sensory attributes of chapatti were evaluated by a panel of judges. The characteristics evaluated and the maximum score possible for each were: height on puffing, 10; pliability, 10; appearance, 20; hand feel, 10; texture, 25; and taste, 25 so that the maximum total score would equal 100. The pliability of chapatti was also determined objectively by the method standardized by Rao (1982) using a specially designed apparatus. The texture was measured with the Warner Bratzler shear press after folding the chapatti into four layers; shear value was recorded in pounds. A chapatti scoring a score of above 70 out of total score of 100 was considered of good quality.

6. **Dhaliwal *et al*** (1996) milled Canadian wheat into flour of 80 per cent extraction rate. Chapatti dough was prepared in a GRL mixer using water absorption and mixing time as derived from 500 BU farinograph data. The dough was allowed

to rest for 15 min before sheeting. A portion of the dough was rounded into a ball and then placed on a rolling board fitted with a thickness guide of 1mm. The ball was rolled using a roller pin in one direction, inverted and then rolled in a perpendicular direction. A stainless steel cutting template of 15 cm diameter was pressed on to the dough sheet to get a uniform diameter and 1 mm thickness circular disc or raw chapatti. The disc was immediately placed on a preheated (225°C) griddle, baked for 40 sec on one side and 70 sec on the other, followed by puffing in an oven preheated at 275°C for 20 sec.

7. The earlier studies have shown that the wheat variety and processing conditions are the main causes of variation in the quality of chapatti. **Navnidhi et al** (2007) further investigated the effects of the processing conditions in order to standardize the process for chapatti making. Medium hard wheat variety C 306, which is known for its suitability for chapatti making, was ground in a laboratory grinder after conditioning to 14 per cent moisture. The wheat flour contained moisture 14 per cent, protein 12 per cent and ash 1.4 per cent, and had farinograph water absorption capacity of 90.6 per cent which was rather on the higher side as compared to the above discussed test baking methods. The dough of optimum quality was prepared in farinograph by mixing 100 g of flour with 2 per cent salt using water equal to water absorption capacity of flour determined in farinograph (90.6 per cent) for 4 minutes. After resting for 40 minutes the dough was sheeted to a uniform thickness of 2 mm using a dough sheeting machine (Model SF 66). The dough sheet was cut into a 15 cm diameter circular disc using a sharp edged circular die. The circular dough disc was baked at 210°C on a thermostatically controlled chapatti baking plate with cut off temperature of 280°C. Baking was continued for 40 and 100 seconds for side one and side two, respectively. The baked chapatti was immediately placed on gas flame having temperature between 290 - 320°C and allowed to puff fully for optimum period of 25 seconds. The height of the puffed chapatti in centimeters was immediately measured using a specifically designed system having a ruler attached to it for convenience. The optimum puffing height was 6.5 cm. The chapatti was then placed on a muslin cloth and cooled at room temperature for 15 minutes. The softness of chapatti was measured using texture analyzer; a firmness force of 248 g was optimum. These optimized conditions for chapatti making process resulted in chapatti of good quality with respect to sensory characteristics.

The chapatti test-baking methods delineated above made use of cast iron griddle (*tawa*) or hot plate with manual temperature control system to bake chapattis. Austin and Ram (1071), however, used hot iron plate kept in a rotary hearth baking oven which was regulated to maintain a temperature of 232.2°C Ebeler and Walker (1983) recommended the use of electric convection oven maintained at 232° C to bake chapattis for laboratory evaluation because it gave uniform reproducible sample.

Mechanization of Chapatti Making

The developments in mechanized chapatti production have progressed during the last two decades or so in view of the increasing marketing potential witnessed for unit packed chapattis. The driving forces have been the changes in lifestyle, occupation pattern and urbanization, which have in turn contributed to changes in culinary practices and dietary habits resulting in increased need for safe and convenient foods. The institutional, industrial, hospitals and armed forces catering establishments, *langar* or *preetibhoj* (sacred meal) serving religious establishments and other mass feeding establishments that require large volume of chapattis and related products have benefited the most from these cost effective mechanized batch or continuous systems of different capacities. These systems are hygienic, overcome human drudgery and large labour force requirements. The chapattis produced on these mechanized systems are hygienic and safe since chapattis are not touched by human hands during processing (Gupta, 2001).

The mechanized chapatti making process involves following operations:

1. Conversion of flour into dough
2. Pressing or Rolling / Sheeting of dough and die cutting to make chapatti dough discs
3. Recycling of dough sheet trimmings
4. Baking cum puffing of chapatti dough discs.

Preparation of Dough

Conversion of whole wheat flour into dough may be effected in dough mixers / kneaders already developed ever since bread making was mechanized. Planetary mixers, sigma blade mixers and conical screw mixers are commonly used. However, the requirement for water to obtain dough of optimum consistency needs to be determined using already developed Brabender Farinograph or Research Water Absorption Meter (Shurpalekar and Prabhavathi, 1976; Rao et al, 1986a) because whole wheat flour used for chapatti making generally requires more water to make dough of optimum consistency as compared to all purpose flour used for bread making. Moreover, the whole wheat flour derived from different varieties of wheat and milled on different mills, requires different amount of water to make dough of desired consistency (Rao et al, 1986a).

Formation of Chapatti Dough Discs (Raw Chapatti)

Flattening and sheeting of the dough is one of the most critical operations in flat bread production, and small variations in thickness change the flat bread quality significantly (Rubenthaler and Faridi, 1982) as discussed above. Following systems / techniques have been developed from time to time either as batch operations or as continuous processes:

1. Pressing or stamping operations
2. Sheeting operations:
   (a) Sheeting between set of rollers and die cutting;
   (b) Sheeting by extrusion and die cutting
3. Two systems have been developed to sheet the dough by extrusion process:
   (a) Pneumatic Pressure Extrusion(Venkateshmurthy and Jayaprakashan, 1991)
   (b) Mechanical Extrusion (Sridhar, 1991)

Pressing or stamping operations

These machines comprise a set of plates of which one moving plate is pressed manually against the stationary plate. The dough ball after slightly pressing between the palms of hands is placed on the stationary plate naked after applying little edible oil or clad with thin polythene sheet which has been smeared with edible oil. Pressing or stamping mechanism varies from machine to machine.

In the machine based on **Indian Standards Institute** standard (IS 9246, 1979) pressure is applied with the help of lever to which the two circular plates are hinged at one end. The chapatti formed is reported to be reasonably round, of uniform thickness and approximately of the same size as that of the working surfaces.

Kannur et al (1985) developed a simple mechanical device for flattening dough into chapatti by modifying a round die flanging machine (M/s Metal Box Company). Two metal plates were attached to the flanger, of which the lower plate is slightly bigger and attached to the lever mechanism. The upper plate is fixed. The dough ball of required weight is placed on a plywood sheet of suitable thickness packed in a polythene pouch and similar plywood sheet is placed on the dough ball. When hand lever is operated the lower plate goes up and the dough is stamped between the two plates giving a flattened dough disc. The chapattis thus prepared are claimed to be of uniform thickness and round in shape and there is no difference in baking characteristics of chapattis rolled by hand and those flattened by this mechanical device.

Bedekar (2001) reported pressing of chapatti dough on an automatic press unit driven by a hydraulic power pack which is operated manually using a pedal. The operator places the dough ball on the lower moving plate which is rammed against the fixed upper plate by actuating the hydraulic cylinder resulting flattening of the dough ball into a disc of desired diameter and thickness.

Gurushree et al (2011) designed and fabricated a portable and manually operated chapatti press cum vermicelli extruder device. The device consists of three steel plates mounted one above the other on a frame. The upper and the bottom plates are fixed while the middle one is movable A screw rod with handle is mounted above through the top plate. The dough piece is placed between two polythene sheets (greased with vegetable oil to prevent dough sheet stickiness) and kept in between middle and bottom plates. The middle plate is pressed with the screw rod driven with the help of attached handle. The dough is pressed to 1.5 mm thick sheet which is then cut into a circle (15 cm diameter) using a steel cutter. The press time per chapatti is 12 s compared to 29 s of manual rolling. The overall sensory quality score of 50.15 and 48.4 for pressed and rolled chapattis respectively showed that quality of chapattis was not adversely affected as a result of mechanical pressing. For making vermicelli, the bottom plate plug of the machine is removed and extruder socket is inserted.

2. Sheeting Between Set of Rollers

Raghavan et al (1994) reported the use of pizza making machine (Pizza Sprint, model M-40, Rheon Co., Japan) consisting two pairs of Teflon rolls to make chapatti dough discs. A thick sheet of dough is first prepared manually and then passed repeatedly through the pairs of rollers to get the desired thickness. The authors studied the effect of key factors such as shape and size of the initial dough feed, roller clearance and ingredients on mechanical sheeting of chapatti dough made from whole wheat flour and resultant *atta*. However, the capacity of this machine is small and the operation is not continuous, as the initial dividing and flattening of the dough have to be done manually (Venkatesh Murthy et al, 1998).

In his schematic diagrams of chapatti makers, Bedekar (2001) has shown the use of series of pairs of rollers to sheet the chapatti dough after passing through hopper device. The sheet then passes under the cutter roller and is cut into the discs of desired shape. The scrap is collected and recycled. In a modified unit there is no hopper used. Instead the chapatti dough is directly fed to a pair of heavy rollers which have provision for reversible dough brake for folding the sheet after applying oil. The folded chapatti sheet is then fed to the main chapatti cutting and baking unit. The folded sheet is first passed through two pairs of rollers to achieve the desired thickness. The dough sheet then passes under the cutter roller to get dough discs of desired shape.

Sheeting by Pneumatic Pressure Extrusion

A batch type machine based on the principle of sheeting by pneumatic extrusion and cutting was developed at Central Food Technological Research Institute, Mysore, India (Venkateshmurthy and Jayaprakashan, NS-1/91-PAT; Venkateshmurthy and Jayaprakashan, NS-4/91-PAT). The machine consists of a conical vessel having a slit opening at the bottom with a provision to vary the gap, and a gasketed lid at the top. The cutter conveyor is placed below the conical vessel, which carries the sheeted dough. The Teflon-coated roller situated at the other end of the conveyor chain facilitates cutting of the sheet when it comes between the conveyor and the roller. There are two dusting units to dust the extruded sheet with flour on both the sides. The return cloth conveyor provided next to the Teflon roller takes out the scrap dough. The chutes are used to divert the circular chapatti discs on to the collection tray (Venkatesh Murthy et al, 1998). The sheeting characteristics of chapatti dough in this machine, as influenced by the type of flour, consistency of the dough and optional ingredients like salt and fat were studied. It was found that the water for making dough must be reduced at least 2 to 3 per cent as compared to the water normally used for manual sheeting of the dough to check stickiness. The slit width of the extruder had to be maintained nearly half the desired thickness of the chapatti sheet. Incorporation of salt considerably improved the sheeting characteristics. The quality of chapatti made using this machine was comparable to manually sheeted chapatti. The changes in the rheological characteristics as well as sheeting characteristics as affected by optional ingredients were more pronounced in the case of resultant *atta* compared to whole wheat flour.

Sheeting by Mechanical Extrusion

The dough sheeting systems discussed above have limitations being batch wise, and are beset with the problem of dough stickiness and non-uniformity in the thickness of the dough sheet (Sridhar and Manohar, 2001). To overcome these problems, a system based on mechanical extrusion was designed and developed at the Central Food Technological Research Institute, Mysore, India for the continuous production of chapattis (Sridhar, 1991).The system powered by 4 h. p. AC motor consists of a barrel, a screw that rotates inside the barrel, a rectangular slit disc, a rotary cutter, a flour duster to dust flour on the dough sheet to control stickiness of the dough and a mechanical volumetric feeder to ensure uniform and continuous feeding of the dough. The extruded dough leaves the die as sheet and passes to a set of rotary cutters mounted on a roller rotating in the direction of moving dough sheet. The forward movement of the dough sheet and the cutter roller rotations are synchronized to avoid shrinkage and tearing of the sheet and to allow recycling of the trimmings.

Developments in Chapatti Making Machines

Based on the above discussed systems for dough flattening / sheeting, a large number of chapatti making machines operating batch wise, semi-automatic or fully automatic and continuous type have been developed by different machinery manufacturers in India and overseas. These have eased out the cumbersome and laborious process of chapatti making not only at household level but also for hotels, institutional / industrial / railway / hospital canteens, and other mass feeding establishments.

Household Roti Makers: A wide range of semi-automatic home Roti Makers have been developed based on the system similar to the **Indian Standards Institute** standard (IS 9246, 1979) and are available in the market in different brands. A Roti Maker is essentially made up of two non-stick coating stainless steel plates (15 to 20 cm diameter) of suitable thickness. The unit is provided with a lever handle for pressing the dough ball into chapatti dough disc, and knobs to lift up the upper plate whenever required. The plates are equipped to heat up electrically. The dough of desired consistency is prepared and allowed to rest for 10 to 20 min and made into dough balls of suitable size and weight. The unit is switched on and the dough ball after slightly flattening between the palms of hands is placed at one end of the lower plate. The upper plate is then lowered over the ball and pressed with the help of lever handle to flatten the dough ball into a circular chapatti dough disc having uniform thickness. The upper plate is then lifted up .The chapatti is baked by turning upside down a couple of times. When bubbles start appearing on the surface of the chapatti, the upper plate is gently lowered on the baking chapatti to promote uniform puffing. There are some fully automatic household chapatti / roti making machines that have built in provision for dough kneading, dough ball making and flattening, baking and puffing operations. In these units required quantities of whole wheat flour, oil and water are placed in their respective containers located at the top of the units. The machines are operated electronically for different operations. These semiautomatic and fully automatic domestic *roti* makers are versatile, easy to use and make chapattis as good as traditionally prepared chapattis.

***Commercial Roti /Chapatti* Makers:** Bedekar (2001) has presented schematic diagrams of three different designs of automatic and continuous chapatti makers, and one design of a semi-automatic machine in his article entitled “Chapati- Indian Bakery Product of the 21st Century” appeared in Indian Food Industry publication. In semi-automatic machine, the chapatti dough discs are produced by pressing on an automatic press unit driven by a hydraulic power pack mentioned above. Two such units are required to feed the continuous slat type conveyorized oven. This machine has capacity to make about 450 to 500 chapattis per hour. Amung the automatic machines, there are three types of designs. **Design No. I** is continuous type with capacity to produce about 1000 to 1200 chapattis per hour. The baking is accomplished on a straight conveyorized oven and chapattis are turned upside down manually.

Design No. II is a continuous automatic chapatti making and baking machine. The chapattis are shown as square or rectangular which produce very little scrap as compared to circular one. This machine has the capacity to produce 2000 chapattis per hour.

Design No. III is also a continuous type chapatti making and baking unit. This unit is equipped with a reversible dough brake for turning dough into continuous sheet which is folded after application of edible oil to prevent sticking. This folded sheet is then fed to the main machine for further sheeting and cutting into the desired shape. The chapatti coming out of this machine has more oil content and remains soft for a longer time. This unit also produces 2000 chapattis per hour. The capacity of these units can be doubled by using double line cutter system. Based on this concept and design, M/S Om Engineering Works, Rajpura (Punjab), India was the first to fabricate automatic chapatti making machine in the year1985-86 (Bedekar, 2001). Since then they have upgraded their technology. Many more semi- and fully automatic machines have been developed using these designs by several fabricators in the country.

***Semi-automatic and Fully Automatic Commercial Machines*:** Besides the use of separate dough kneaders for preparation of dough of desired consistency, dough ball making, flattening / sheeting systems are separated from the baking cum puffing systems in semi-automatic machines whereas fully automatic machines have these systems placed one after the other, i.e. in continuous framework as shown by Bedekar (2001).

***Dough Pressing Type Machines*:** In this type of machine the dough balls are either prepared manually or by using a separate dough ball making machine. The dough balls are fed to the flattening unit one by one manually and the flattened dough disc is carried to the baking cum puffing oven on a conveyer system. Some compact chapatti making machines are fitted with a dough ball making unit followed by a flattening unit. The dough is fed to the dough ball making machine through a hopper fitted with a feeding device. The ball formation and flattening operations are synchronized. In some chapatti pressing type machines (e.g. **rotary indexing type**), the dough ball is manually placed on the revolving circular plate and flattened by stamping mechanism. The flattened dough disc as it reaches the operator; it is manually transferred to the lower revolving heated circular plate to bake and puff as it goes round before reaching the outlet for the baked chapattis. In another pressing type of machine, the balls are made in a dough ball maker. The balls then manually fed to the flattening unit. The flattening unit may be stamping / pressing type or set of rollers to flatten the dough ball. The flattened chapatti dough disc directly drops through a chute on to the set of three to four rotating circular plates fitted one above the other vertically and heated by LPG. The baking chapatti moves from the upper plate to the lower plates in sequence and comes out as fully puffed chapatti. In another fully automatic type chapatti making machine the ball cutting unit, pressing unit and the baking unit are integrated. The flattened raw chapattis are transported on a conveyor belt to the rotary *tawa* system to bake the chapattis, which simulate the traditional home like *tawa*-baked-chapattis.There are many more designs of pressing type of chapatti machines being marketed.

***Dough Sheeting Type Machines*:** The dough sheeting may be accomplished in this type of machines using different devices namely set of rollers, pneumatic pressure extrusion system or mechanical extrusion system as stated above.

Sidhu et al (1987) reported semi-mechanized production of chapatti using a dough kneader, dough ball maker, a dough ball rolling system to get chapatti dough discs of required diameter and thickness. The chapatti dough discs were manually loaded on to the conveyer for baking cum puffing in the oven. This and other systems developed earlier had limitations as mentioned above.

Central Food Technological Research Institute (CFTRI), Mysore (India) has developed a prototype design of fully automatic chapatti making machine using an indigenously developed mechanical extrusion system to sheet the dough continuously. In this machine, the wheat flour dough prepared in a planetary mixer is conveyed continuously to the extruder. A die directs the extruded dough sheet to the endless belt conveyor. The sheet while travelling on the conveyor passes through a rotary cutter and gets uniformly cut in to circular shapes. These circular raw chapattis then enter the baking chamber while the trimmings / cut portions of the sheet are trapped and sent back to the extruder. LPG is used as source of heat to bake the chapattis. The baked and fully puffed chapattis can be served hot or packaged after cooling. The design, drawings and knowhow of this continuous chapatti making machine has been released to several entrepreneurs for production.

Sridhar and Manohar (2001) at this institute have standardized the above chapatti making process using the continuous mechanical extrusion system described above to optimize the processing variables such as size of the die opening, production rate and initial moisture content. Based on the response surface analysis for minimum specific energy and maximum puffing index and sensory analysis of chapatti, the optimum conditions were found to be: a rate of 40 kg/h dough, die opening of 1.4 mm height x 200 mm width and dough moisture contents of 60 and 75 per cent (w.b.) for resultant *atta* and whole wheat flour, respectively. The sensory analysis showed that the quality of chapattis produced under these optimized conditions compared well with traditionally prepared chapattis.The main advantages of this system are: a) simplicity of design, b) higher rate of production with uniform quality, c) improved hygiene, d) low maintenance, and e) inexpensive operation. The process flow chart giving briefs of the above system to produce well-puffed chapattis is shown in **Figure 2.**

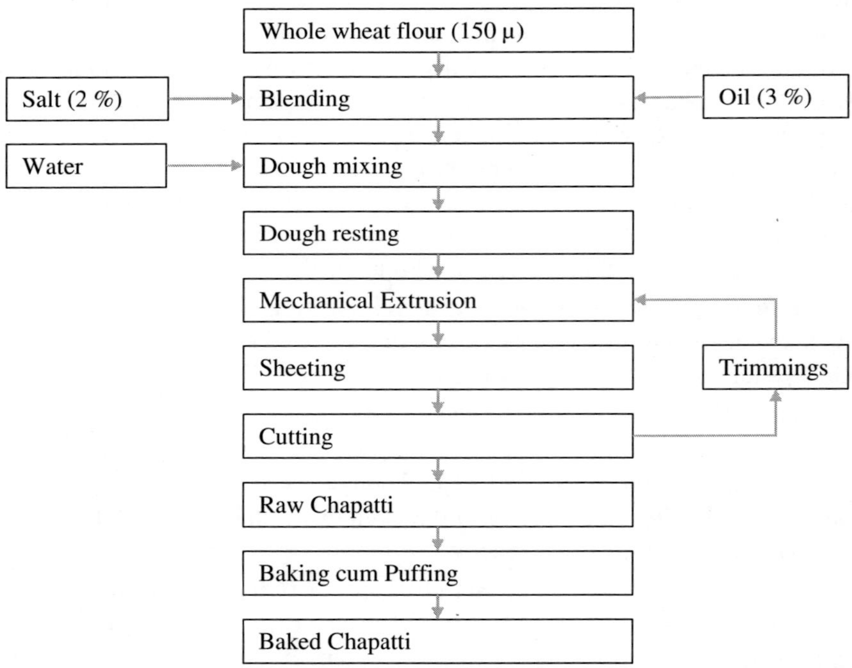

Figure 2: Process flow sheet for continuous production of chapatti
(**Source:** Sridhar and Manohar, 2001).

Currently, there are several machinery manufacturers in India and overseas who have ventured into fabricating chapatti making and baking machines of semi-automatic, fully automatic and continuous type with different designs and variable capacities. There is huge potential in the business of commercial preparation and marketing of chapattis and related products, Manufacturers and suppliers of these machines may be contacted through Trade India online portal for further details.

Chapter 5

Nutritive Value of Chapatti

Chapatti and related products constitute an important food source of energy and several vital nutrients such as proteins, minerals, and vitamins. The nutritive value of chapatti is, however, the reflection of the chemical composition of the wheat used as only whole wheat flour and water are generally used to make chapattis. The wheat used should, therefore, possess such nutritional quality traits as high protein content with well-balanced levels of amino acids especially lysine, bioavailable micronutrients, and a suitable starch composition (Ram and Mishra, 2007).

Wheat is essentially composed of protein, carbohydrates, dietary fibre, fats and lipids, minerals, vitamins, and other minor components such as phytonutients and antintrients. The chemical composition of whole-grain wheat flour (**Table 1**) and nutritional value of chapatti (**Table 2**) are given below:

Table 1: Chemical composition (per 100g) of whole-grain wheat flour

Proximate	Value (g)	Vitamins	Value (mg)	Minerals	Value (mg)
Moisture	10.74	Thiamin	0.50	Calcium	34.00
Protein	13.21	Riboflavin	0.17	Iron	3.60
Total lipids	2.50	Niacin	4.96	Magnesium	137.00
Ash	1.58	Pantothenic acid	0.60	Phosphorus	357.00
Carbohydrates (by difference)	71.97	Pyridoxine	0.41	Potassium	363.00
Total dietary fibre	10.70	Total folate	44.00 µg	Sodium	2.00
Total sugars	0.41	Choline	31.20	Zinc	2.60
Starch	57.77	β-carotene	5.00 µg	Copper	0.41
		Vitamin A	9.00 IU	Manganese	4.07
		Lutein and zeaxanthine	220.00 µg	Selenium	61.80 µg
		Tocopherols (α, β, γ)	2.85		
		Betane	72.80		

Source: USDA Database (2016).

Table 2: Nutritional value of chapatti (per 100 g).

Proximate	**Value** (g)	**Minerals**	**Value** (mg)	**Vitamins**	**Value** (mg)
Protein	11.25	Calcium	93	Thiamin	0.55
Total lipids	7.45	Iron	3.01	Riboflavin	0.2
Total fibre	4.9	Magnesium	62	Niacin	6.78
Total sugars	2.72	Phosphorus	184	Pyridoxine	0.27
Carbohydrates	46.36	Potassium	266	Vitamin E	0.88
		Sodium	409		
		Zinc	1.57		

Source: USDA Database (2016).

Protein

Protein is the primary nutritive component of wheat and whole wheat flour. There are two types of wheat proteins- gluten forming (85 per cent) and non-gluten forming (15 per cent). Gluten forming proteins comprise low molecular weight gliadins and high molecular weight glutenins. Non gluten forming proteins comprise albumins and globulins which are water and salt soluble, respectively. Comparatively, gliadins and glutenins are deficient in lysine, whereas albumins and globulins are rich in lysine (Pomeranz, 1988).

Protein is considered the most important nutrient for human and animals. Wheat with high protein content is preferred because wheat and other cereals form the bulk of the ordinary Indian diet and provide more than 66 per cent of the daily protein requirement of an individual (Juneja et al, 1980).The primary function of protein in the diet is promotion and maintenance of growth. Though wheat constitutes between 8 to 15 per cent protein (Austin and Ram.1971) but its quality in terms of the amount and relative proportion of different amino acids is unfortunately not well balanced as required by man for proper growth. The deficiency of proteins causes serious nutritional disorders in both infants and adults. It is, therefore, necessary that the diet must contain proteins which are sufficient not only in quantity but also in quality. The quality of protein is determined by the amount and relative proportion of the different amino acids contained in it, and digestibility and bioavailability of amino acids. There are several methods (FAO/WHO, 1991) used to determine the protein quality such as biological value (BV), net protein utilization (NPU), chemical score, protein efficiency ratio (PER) and protein digestibility corrected amino acid score (PDCAAS).

There are 19 amino acids that occur in proteins. Some of these amino acids are called essential amino acids (**EAAs**) as these cannot be synthesized in the body and the food consumed must necessarily contain these in adequate quantity. There are eight **EAAs** including isoleucine, leucine, lysine, methionine, phenylalanine, threonine, tryptophan and valine, each of which must be in adequate quantity for complete utilization of the protein by the body. The protein containing all the EAAs is considered complete, having high biological value (BV) and high net protein utilization (NPU). Most plant proteins especially cereal proteins are nutritionally incomplete due to the deficiency of one or more of these **EAAs.** The deficiency in a certain amino acid reduces the availability of others present in abundance. Wheat protein is deficient in lysine, tryptophan, methionine and threonine (**Table 3**). Because of this deficiency, these **EAAs** become the limiting amino acids in wheat and whole wheat flour. Though protein digestibility is higher in white flour, overall amino acids availability is higher in whole wheat flour (Ram and Mishra, 2010). Nevertheless, supplementation with flours of other food materials, e.g. other cereals and legumes or oilseed meals which are rich in these **EAAs,** may be practiced to enhance the nutritive value of whole wheat flour. Alternatively, the flour may be suitably fortified with one or more of the limiting amino acids. Higher intake of lysine than that found in wheat is particularly essential for children for growth. The need for lysine is about three times more for children than adults.

Table 3: Amino acid composition (g/100g) of whole-grain wheat flour

Amino acid	Value	Essential amino acid	Value
Alanine	0.489	Isoleucine	0.443
Arginine	0.648	Leucine	0.898
Aspartic acid	0.722	Lysine*	0.359
Cysteine	0.275	Methionine*	0.228
Glutamic acid	4.328	Phenylalanine	0.682
Glycine	0.569	Threonine*	0.367
Histidine	0.357	Tryptophan*	0.174
Proline	2.075	Valine	0.564
Serine	0.620		
Tyrosine	0.275		

Source: USDA Database (2016). *Limiting amino acid

Another limitation of wheat proteins is associated with poor digestion and allergic reactions. Bakers' asthma is a typical condition in which water soluble flour proteins- albumins and globulins are implicated as a result of inhalation of flour particles. Celiac disease or gluten sensitivity is a condition that results in damage to the small intestine resulting in malabsorption of nutrients. The common symptoms are poor growth, diarrhea and abdominal pain. The gluten proteins- gliadins and glutenins are implicated as the major causative agents of celiac disease (Shewry, 2009).

Carbohydrates

Carbohydrates constitute the bulk of the total dry matter of the wheat grain and normally occur as monomers and polymers, and include sugars, oligosaccharides and the polysaccharides. Sugars comprise monosaccharides (e.g. glucose, fructose, galactose) and disaccharides (e.g. sucrose, maltose), which occur in lesser amounts located primarily in the bran and the germ of the wheat grain. The oligosaccharides comprise raffinose and stachyose. The polysaccharides comprise starch and non-starch polysaccharides.

Wheat grain may contain starch to the extent of 60 to 75 per cent of the grain dry matter. Starch occurs in the wheat grain in the form of granules-small, spherical in shape and large, lenticular in shape. Small spherical granules dominate. When starch is heated with excess water, the granules swell resulting in gelatinization which alters the molecular structure with in the granules. The concentrated suspension of gelatinzed starch when cooled, retrogradation takes place leading to molecular reassociation and crystallization followed by gel formation. The rheological properties of the starch gel influence the texture and acceptability of the wheat flour products.

Chemically starch is basically a polymer of glucose. Amylose (linear) and amylopectin (branched) are the two main polymers constituting the starch in wheat grain and the whole wheat flour. Normal wheat starch is composed of 20 to 30 per cent amylose and 70 to 80 per cent amylopectin (Sramkova et al, 2009). Amylose to amylopectin ratio affects the physical properties of starch. The two play different roles in gel formation. The short term retrogradation of starch gels is due to gelation and crystallization of solubilized amylose fraction where as long term retrogradation of starch gels is attributed to gelation and recrystalization of the amylopectin fraction (karim et al, 2000)..

Starch is the main source of energy in the wheat based diets. Chapatti is reported to contain high amount of slowly digestible starch and lesser amount of rapidly digestible starch. However, starch digestibility in chapatti is influenced by the accompaniments rather than the wheat base (Sharavathy et al, 2001). Chapatti is normally consumed along with vegetarian or non-vegetarian accompaniments.

The non-starch polysaccharides comprise cellulose, pentosans or arabinoxylans and β-glucans. They together make up the bulk of the dietary fibre, which also includes lignin plus resistant oligosaccharides and resistant starch. Pentosans are the major non-starch polysaccharides in wheat and consist of xylan backbone with branches of arabinose residues in various linkages. Pentosans play an important role in water balance of wheat dough, rheological properties of dough, retrogradation of starch, and baked products making quality (Izydorczyk and Biliaderis, 1995). Pentosans have been reported to play a significant role in the chapatti making quality of wheat. Revanappa et al (2007) have shown that higher the arabinose / xylose and xylose / glucose ratios in the whole wheat flour better the chapatti quality in terms of overall sensory scores. A comparative view of chapatti making quality of some selected wheat varieties with arabinose / xylose ratios is given in **Table 4**. This suggests that variation in structural features of pentosans could be responsible for the differences in chapatti-making quality of different wheat varieties.

Table 4: Comparison of chapatti making quality in wheat varieties with arabinose / xylose (Ara/Xyl) ratios

Wheat variety	Overall quality (50)	Ara / Xyl ratio
GW 322	41.7	1.32
DWR 162	42.0	1.26
Sonalika	Good	1.25
HD 2189	36.0	0.84
MACS 2496	35.2	0.80

Source: Revanappa et al (2007); Nandini and Salimath (2001).

The carbohydrates in whole wheat flour are not only the excellent source of energy but also the dietary fibre. Though the amount of dietary fibre in whole wheat flour is small, consumption of large amount of flour in the form of chapattis contributes to significant amounts in the diet. The defatted whole wheat flour of the variety Sonalika has been reported to contain soluble and insoluble dietary fibre to the extent of 1.0 and 10 per cent, respectively (Nandini and Salimath, 2001).The fibre helps prevent constipation and may also protect against diseases such as diverticulitis, hemorrhoids, and colon cancer (Plaami, 1997). The benefit of higher intake of fibre from whole wheat flour may, however, be counterbalanced by the phytate content of the wheat bran. Phytate binds minerals including iron, zinc, magnesium and calcium by forming insoluble compounds that reduce bioavailability (Kumar et al, 2010).

Lipids

Wheat grain contains 2 to 4 per cent lipids, majority of which are located in the germ and the bran. The amount of lipids in wheat flours may vary with the extraction rate; the whole wheat flour containing the highest amount. Lipids exist in wheat flour as starch lipids or bound lipids, and as non-starch lipids or free lipids. Starch lipids are amylose-complexed monoacyls while non-starch lipids are triglycerides, diacylglycerides and free fatty acids. Starch lipids are effectively inert and non-starch lipids participate in physical, chemical and biochemical processes in the flour and the dough (Morrison, 1978). The bound lipids are mostly phospholipids and glycolipids. The typical fatty acid composition of wheat comprises linoleic, oleic, palmitic and linolenic acids. The unsaponifiable components of lipids in wheat are the tocols and carotenoids plus sterols. Among the sterols, β-sitosterol and compesterol are the most abundant.

Besides their nutritional importance as source of polyunsaturated fatty acids and fat soluble vitamins, lipids in wheat are also implicated in ageing of wheat flour, mixing behavior of dough, baking and freshness retention in baked products (Arya et al, 1978).

They have a significant effect on the quality and the texture of wheat flour based food products because of their ability to associate with proteins due to their amphipathic nature and with starch forming inclusion complexes. Lipids play a significant role in baking of flour based products. They modify the gluten structure at the dough mixing stage; catalyze the oxidation of SH groups in proteins and polymerization of proteins through lipid peroxidation. They prevent interactions between starch granules during gelatinization. They retard water transport from protein to starch thus acting as antistaling agent (Pomeranz, 1988). During dough mixing and baking the major portion of free lipids get bound to proteins and starch. Comparatively polar lipids, phospho- and galacto-lipids are bound to larger extent than neutral lipids. Removal of petroleum ether soluble lipids from *atta* does not influence the texture 'hardness' of chapattis but removal of total lipids adversely affects the texture of chapattis The major portion of phospho- and galacto- lipids become inextricable during chapatti preparation and wheat varieties having higher levels of polar lipids yield chapattis that retain relatively softer texture during storage (Arya et al, 1978). Wheat flour lipids also appear to contribute to the flavour of chapatti by augmenting the carbonyl composition of chapattis through not only generating a large number of additional aldehydes, but also retaining the volatile carbonyls formed by other reactions by exerting a solvent action (Arya et al, 1976).

Minerals

The major constituents of the mineral fraction of whole wheat are calcium, phosphorus, magnesium and potassium with significant amounts of iron, copper, zinc and manganese (**Table 1**). Wheat is also an important source of selenium. These minerals are largely concentrated in the wheat bran. The milling process may result in significant losses of minerals especially copper, zinc, iron and manganese. Whole wheat though rich in phosphorus but most of it is biologically unavailable to humans because it is mostly present as phytate of calcium, magnesium and potassium located primarily in the bran. The whole wheat flour of 100 per cent extraction which is commonly used for making chapattis contains all these minerals. Nevertheless, their bioavailability is inhibited because of chelation with wheat phytic acid forming insoluble complexes (Kumar et al, 2010).

Minerals play a very prominent role in human nutrition (FAO/WHO, 2004). The formation and maintenance of bones and teeth require calcium, magnesium and phosphorus. The functioning of body cells, blood, muscles, etc. needs the presence of iron, calcium, phosphorus and magnesium. Copper and iron are involved in the formation of haemoglobin and myoglobin preventing anemia. Iron plays a critical role in electron transport and DNA synthesis for all cells, in the synthesis of some hormones and connective tissue. Inadequate amount of iron in the diet is linked to not only anemia but also leads to impaired physical and mental development, decreased capacity to work, reproductive problems, etc. The soluble salts of sodium, potassium, phosphorus and chlorine compose the body fluids maintaining the osmotic pressure as well as acid-base balance. Potassium, zinc and copper function as activators of enzymic systems.

Zinc improves intestinal absorption and is involved in fertility and sterility aspects of reproduction. Zinc deficiency not only impairs growth but also contributes to diarrhea and pneumonia and malaria related complications among young children. Selenium is an essential micronutrient for humans with its antioxidant, anti-cancer and antiviral properties, and provides protection against cardiovascular diseases and boosts immune function (Arthur, 1999).

Vitamins

The vitamins in whole wheat are largely concentrated in the germ and bran fractions of the grain. Milling wheat into flour results in significant loss of certain vitamins the extent depending upon the extraction rate. Fat soluble vitamins- A, E, D and K occur only to a very small extent in wheat because of the low levels of lipids in the grain. Notwithstanding, the germ fraction of the wheat grain contains relatively high levels of lipids and is rich in α- and β-tocopherols and α- and β-tocotrienols with vitamin E activity. Among the cereals, the wheat tocols are considered to possess particularly good vitamin E activity and antioxidant properties (Morrison, 1978). Vitamin E inhibits lipid oxidation, thereby protecting cell membranes and other cellular components. Wheat is not a significant source of vitamin A precursors having any nutritional significance. The major carotenoids of bread wheat, however, are lutein, β-carotene and β-cryptoxanthin (**Table 1**).

Whole wheat is, on the other hand, a good source of B-complex vitamins especially thiamin, riboflavin, niacin, pantothenic acid and pyridoxine with a moderate amount of folate (**Table1**).The B-complex vitamins are associated with carbohydrate, protein and fat metabolism and release of energy. Pyridoxine is a powerful antioxidant and plays a role in the prevention of a number of chronic diseases. The deficiency of these vitamins causes several deficiency diseases, e.g. *beriberi,* pellagra, nutritional diarrhea, etc. Folate is associated with synthesis of nucleoproteins and choline, and prevents anemia. The deficiency of folate may increase the risk of neural- tube birth defects in the baby (FAO/WHO, 2004).

Polyphenolics and Related Components

Apart from supplying a considerable amount of calories, proteins and carbohydrates, whole wheat also provides a range of other components cumulatively designated as phytochemicals. These are phytoactive non-nutritional compounds with diverse positive effects on human health as they are thought to be implicated in protection against several chronic diseases (Thompson, 1994). The phytochemicals found in wheat grain include phenolics, carotenoids and tocochromanols most of which are concentrated in the germ and bran. In whole wheat flour, the bran and germ fractions contribute a large portion of the total phenolics, flavonoids, and carotenoids contents (Liu, 2007). Carotenoids comprise zeaxanthin, lutein, β-carotene and β-cryptoxanthin. Phenolics include phenolic acids, flavanoids and lignans. The phenolic acids of significance are ferulic, vanillic, p-coumaric, protocatachuic, syringic, p-hydroxybenzoic, caffeic and genitisic acids. Ferulic acid is most abundant in whole wheat and is a very good antioxidant (Rice-Evans et al, 1997). The flavanoids include flavone tricin, two glycosides of the flavone apegenin with small amounts of flavanolcatechin, and proanthocyanidin (condensed tannins). Sharma et al (2016) have quantified 14 phenolic compounds in two wheat varieties (C306 and Sonalika) differing in chapatti quality. In the poor chapatti quality bread wheat (Sonalika), the concentration of the 14 polyphenolics was higher than that of the good chapatti quality variety C306. Phenolic compounds are reported to influence the quality of bread. It is presumed that they can affect the quality of chapatti as well. Since chapatti is made from whole wheat flour having higher content of bran and germ which are rich in polyphenolics, the contribution of phenolic compounds to chapatti quality parameters such as colour, aroma and taste is therefore anticipated. However, protein-phenolic interactions are reported to affect functional, and nutritional properties of proteins. (Ozdal et al, 2013).

Lignans are one of the major classes of phytoestrogens. They are a group of phenylpropanoids with potential antioxidant and cancer preventing properties. Lignans in whole wheat are mainly concentrated in the germ / bran. Smeds et al (2007) have identified six lignans, e.g. secoisolariciresinol, hydroxymatairesinol, lariciresinol, matairesinol, pinoresinol and syringaresinol in the wheat bran.

Volatile Flavouring Compounds in Baked Chapatti

The freshly baked chapatti possesses a typical wheaty aroma. The carbonyle compounds have been reported to play an important role in the aroma profile of fresh chapatti. The carbonyls that have been identified in freshly baked chapattis include diacetyl, acetoin, glyoxal, formaldehyde, acetaldehyde, propanal, acetone, crotonaldehyde, ethyl methyl ketone, valeraldehyde, isovaleraldehyde, hexanal, nonenal and 2, 4-decadienal (Kannur et al, 1974). These volatiles are possibly generated via non-enzymatic browning reactions, and degradation of unsaturated fatty acids and free amino acids during baking of chapattis (Arya et al, 1976). Further research is required to investigate the role of chapatti dough volatiles and starch plus gluten in the development of typical wheaty aroma through non-enzymatic browning reaction between free amino acids and sugars in the chapatti dough during baking of chapatti as investigated for pan bread (Berry, 1973; Berry and Gramshaw, 1988). The role of wheat bran in the formation of wheaty aroma is needed to be explored.

Factors Affecting Nutritional Quality of Chapatti

Since chapatti is made from whole wheat flour it should be nutritionally superior to white bread because of more protein, minerals, vitamins, and other nutrients in the whole wheat flour compared to the white flour of low extraction used for white bread. On the contrary, early studies have shown that the digestibility of protein and carbohydrates falls, and minerals bioaccessibility decreases as the flour extraction rate increases (Moran and Pace, 1942; McCance and Walsham, 1948). This contradiction could be correlated with the relative proportion and interaction of the different nutrients present in the flours. Moreover, the nutritional quality of cereals is determined not only by their protein and energy content and amino acid composition but also depends upon the methods of processing before consumption, and the presence of factors that affect the digestibility and bioavailability of these nutrients (Lathia and Koch, 1989). Factors that affect the nutritional quality of chapatti are discussed here under.

Effect of Processing

(a) ***Milling*:** Wheat flour extraction rate has marked significant effect on its nutritional profile. The lower extraction rate flours tend to have reduced level of minerals and phytochemicals. With increased extraction rate, the levels of protein, fibre, sugar, lipids, mineral matter and phytochemicals increase whereas starch content decreases. The chapattis prepared from whole wheat flour would, therefore, possess higher level of all the nutrients. Whole wheat flour generally used for chapatti making is normally milled in stone or plate mill in which high temperature ranging between 85 to 95^{0}C is generated during grinding. The heat produced is reported to have a significant effect on lipids, protein fractions, and diastatic and proteolytic activities (Ranga Rao et al, 1986). Free lipids tend to increase whereas bound lipids decrease. Though there is no change in total proteins, salt soluble proteins as well as gliadin increase slightly while high molecular weight glutenin fractions decrease. Prabhasankar and Rao (2001) have shown that heat produced during grinding of wheat in plate/stone mill denatures the protein, and reduces the amino acid content. A decrease in the unsaturated fatty acids content was also observed. This shows that there is slight loss of some essential amino acids and fatty acids upon grinding whole wheat flour in plate/stone mills. The whole wheat flour ground in stone mill lost more than 50 per cent each of lysine and linolenic acid eventually leading to decline in the nutritional value of chapatti. The vitamins thiamin and niacin are also sensitive to temperatures generated during milling of whole wheat flour. The losses of thiamin and niacin are reported to be between about 14 to 55 and 11 to 23 per cent respectively when *atta* is milled in hand-driven *chakki*, water-driven stone mill, power-driven stone mill and roller mill, the lowest losses occurred in hand-driven *chakki* and highest in power-driven stone mill (Austin and Ram, 1971).The resultant *atta* also known as '*mill atta*', a byproduct of roller flour mill is, on the other hand, superior to whole wheat flour nutritionally as it contains about twice as much thiamin and one and a half times as much riboflavin and lysine, respectively as compared to whole wheat flour (Tara et al, 1969; Austin, 1978b). Resultant *atta* being cheaper than whole wheat flour, is generally used for the preparation of chapattis and *tandoori rotis* in restaurants, hotels, *dhaabas* (roadside eateries), and industrial canteens, but hardly in households. Except for bright whitish appearance, the chapattis made from it are inferior in quality as these have bland taste and lack in wheaty aroma and become tough shortly after baking. The food processing operations, e.g. milling of cereals and subsequent baking carried out, either commercially or at the consumer level significantly influence the level and stability of vitamin E in the finished product. Nielsen and Hansen (2008) reported higher vitamin E content in whole wheat flour milled in stone mill compared to roller milled flour. This difference being due to the absence of germ and bran fractions in roller milled flour. The total loss of vitamin E during storage (297 days) was 24 per cent for stone milled flour and 50 per cent for roller milled flour suggesting enhanced antioxidant activity in stone milled flour. Silvestro et al (2014) observed losses in soluble dietary fibre and bound polyphenols during 6 month storage of flours milled in stone water mill and stone mill in which flour attained temperature of 30 and 60^{0}C, respectively. Comparing the milling techniques, stone ground wheat grains showed the highest amylose and resistant starch contents, which could be of interest for lowering glycemic index. The whole wheat flour milled in disc (stone) mill generally contains higher levels of damaged starch with corresponding increased level of reducing sugars. During baking of chapatti reducing sugars undergo non-enzymatic browning reaction with available amino groups resulting in the formation of volatile and non-volatile compounds of which some could be undesirable. Mulla et al (2010) have reported increased acrylamide formation in chapattis and related products with increased damaged starch content in whole wheat flour. Acrylamide is suspected to have carcinogenic properties.

(b) ***Baking*:** Moderate heat treatment may improve the nutritive value whereas excessive heat treatment decreases the nutritive value of foods (Milner and Carpenter, 1969). The most important factors determining the loss in nutritive value are the duration and temperature level during the cooking process and the levels of moisture and reducing substances (Bender, 1972). The literature on the protein quality of wheat flour and yeast breads is in abundance. However, limited reports are available on the protein quality of cooked or baked wheat chapattis. According to Shyamala and Kennedy (1962), the protein quality of the cooked whole wheat flour chapattis, as determined by body weight of rats and protein efficiency ratio (PER) was appreciably better than that of the uncooked whole wheat flour. This increase in PER is contrary to the results obtained when wheat flour is baked into bread. A possible explanation for this increase in PER of the chapattis as compared with the raw whole wheat flour might be the destruction of trypsin inhibitor during baking. Also, the PER can be influenced significantly by palatability effects as cooking of foods improves palatability (Milner and Carpenter, 1969). Matthews et al (1969) reported that lysine fortification (0.1 to 0.6 per cent lysine monohydrochloride) of *atta* had a detrimental effect on the puffed volume of chapattis though baking loss of only about 4 per cent free lysine occurred. Fortification of wheat *atta* with lysine at levels up to 0.4 per cent is feasible, higher levels depress puffed volume and sometimes darken the colour of baked chapattis. The non-enzymatic browning reaction between amino acids and sugars in chapatti dough during baking of chapatti not only results in dullness of chapatti surface and appearance of burnt spots but also causes losses in nutritive value of chapattis (Abrol et al, 1972b). Tara et al (1972) investigated the effect of lysine on protein quality of whole wheat meal chapattis and reported losses in lysine to the extent of 4.5 per cent only compared with leavened bread (12.5 per cent) upon baking. This difference could be attributed to shorter period of baking time for chapattis as compared to bread. Eggum and Duggal (1977) reported slight differences in the amino acid composition of chapatti from those of whole wheat flour. The contents of lysine and tryptophan appear to be most affected and were slightly more than 5 and 10 per cent, respectively lower than those of whole wheat flour. The true digestibility (TD) of chapatti was slightly but significantly higher, whereas biological value and net protein utilization (NPU) were slightly but significantly lower than those of whole wheat flour. Khan and Eggum (1978) examined the nutritive value of Pakistani bread (*roti*) prepared from whole meal wheat flour chemically, by proximate, fatty acid and amino acid analysis and biologically in N-balance experiments with growing rats. Yeast was not used in the preparation of dough. In general, the nutritive value of *roti* was affected only to a minor extent. Bhat and Vivian (1980) while studying the effect of addition of soy, peanut and cottonseed flours on the protein quality of chapattis reported losses in available lysine ranging from 4.7 to 13.8 per cent due to baking. Chawla and Kapoor (1982) while studying the effect of triticale on the quality of wheat chapatti, reported losses in total (5.96 to 8.33 per cent) as well as available (3.89 to 25.11 per cent)lysine upon baking the chapattis. Yasoda and Geervani (1983) reported loss of lysine and tryptophan to the extent of 10.6 and 9.0 per cent respectively during the preparation of *pulka* (well puffed chapatti) from whole wheat flour. The availability of these amino acids was also affected, lysine to the extent of 10 per cent. Longer cooking time, higher cooking temperature and larger proportion of reducing sugars were found to accelerate the destruction of both the amino acids. According to Kaur and Hira (1988) the crude protein, methionine, cystine and available lysine contents of chapatti were lower than those of raw flours, whereas those of *paratha* were even lower than those of chapatti. The fall in cysteine and available lysine in durum wheat *paratha* was as high as 20 and 12 per cent, respectively. This larger decrease in amino acids content of *paratha* compared with chapatti was caused by the longer time and higher temperature of the cooking procedure (shallow frying) required for *paratha.* Anjum et al (2005) reported loss in content of amino acids with consequent decline in amino acid score during baking of whole wheat flour chapattis. The decline in lysine content on 12.4 per cent protein basis ranged between 1.11 to 23.58 per cent in chapattis prepared from different wheat varieties.

A large segment of the population in Indian subcontinent depends upon whole wheat flour chapatti for thiamin intake. Whole wheat flour is a good source of B-complex vitamins as whole wheat grain is milled to produce the flour. Losses in the vitamins during cooking and baking are common. Chaudri and Muller (1972) opined that loss of thiamin during baking of chapatti depends upon the conditions of baking and the thickness of the product. They observed that at a time averaged temperature of 257.5°C, thiamin loss increased from 18.3 to 62.1 per cent when chapatti thickness was reduced by half. There is no uniformity in the chapatti baking parameters such as thickness of dough sheet, time and temperature of baking at household level. In rural areas of the subcontinent thick chapattis are generally eaten, whereas in cities thinner ones are preferred. Thinner chapattis would, therefore, have excessive thiamin losses. Shahidi and Muller (1975) reported that on baking thiamin loss increased almost linearly with decreasing chapatti thickness. Maximum loss was 47 per cent at 1.5 mm thickness and minimum loss 9.8 per cent at 4.1 mm thickness. At the same thickness, variation in baking temperature had a greater effect on thiamin loss than baking time. As the baking chapatti temperature exceeded 100°C, thiamin destruction rose rapidly to 49 per cent. When the chapattis were enriched with synthetic thiamin from the original level of 2.66μg/g to 10.36μg/g, heat destruction decreased with increasing thiamin concentration, confirming that destruction occurs mainly at the chapatti surface. Yadav et al (2008b) standardized the baking parameters of chapatti in respect of thickness (3.2 mm) of dough sheet, time (2.1 min) and temperature (216.3°C) of baking. Under these optimized conditions, the prepared chapattis had greater retention (82 to 85 per cent) of thiamin as compared with those prepared at household level (60 to 70 per cent) under normal conditions. Thus, suggesting that losses in thiamin content may be reduced by adopting the optimized conditions of chapatti baking at household level as well. These authors, however, did not consider the effect of puffing of chapatti on open fire or in preheated (~350°C) oven, which may incur additional losses of the vitamin as the surface temperature is higher though for a very short period of time. The act of chapatti puffing on open flame is commonly exercised at household level. The frozen storage (-18°C) of ready-to-bake chapatti also resulted in decrease in B-complex vitamins, thiamin and riboflavin in particular (Yadav et al, 2009b). Dutta et al (1980) reported that the whole wheat flour contained higher levels of folate compared to the chapattis made from it, suggesting loss of folate during baking.

Effect of Antinutrients Occurring in Wheat

The antinutients are plant constituents that interfere in the absorption and/or utilization, i.e. bioavailability of essential nutrients in the digestive system. Whole cereal grains contain relatively high levels of antinutrients but are deficient or low in substances that promote the bioavailability of these nutrients (Welch and Graham, 1999).Wheat contains several antinutrients such as phytic acid, dietary fibre, protease and amylase inhibitors, lectins, and polyphenolics. These affect the absorption and/or utilization of several minerals, proteins and starches and thus interfering with normal growth, reproduction and well-being, when consumed regularly in amounts occurring in a normal component of the diet. The elimination or reduction of these antinutients from the diet can, however, be achieved either by selection of plant genotype with low level of these factors or through postharvest processing, e.g., extraction, leaching, germination, fermentation, boiling/cooking, etc.

***Effect of Phytic acid*:** Phytic acid known as inositol hexakisphosphate (IP6), a saturated cyclic acid, is the principal storage form of phosphorus in many plant tissues especially of importance in cereals and legumes (Kumar et al, 2010). In wheat phytic acid in the form of phytates, is highly concentrated in the outer layers of the kernel especially bran. It is reported that wheat contains between 40 and 54 percent of the total phosphorus as phytic phosphorus (Mehdi and Abrol, 1972; Swaranjeet et al, 1982; Syed et al, 1992). However, Bains (1949) has shown that phytic phosphorus content of wheat could be varied between 63.0 and 76.6 per cent within the variety depending upon the type of fertilizers used. Austin and Ram (1971) reported phytic phosphorus in whole wheat flour *atta*) from 17 different improved wheat varieties in the range of between 105.55 (C 303) to 146.60 (NP 890) mg per 100g dry matter, and recommended the use of wheat varieties that are low in phytic phosphorus for making chapattis. Kumar and Sushma (2015) recently studied the phosphorus profile of 20 Indian wheat varieties and reported between 77 and 92 per cent total phosphorus other than inorganic phosphorus. The phytic acid content of *chakki atta*, i.e., whole wheat flour (WWF), and that of commercial wheat flour was reported almost the same suggesting that bran is not removed during the commercial milling process (Khan et al, 1986). Partial removal of bran from WWF may reduce the level of phytic acid as shown by Sabir et al (1982) who have indicated that removal of 10 per cent bran alone resulted in a 20 per cent reduction in phytic acid level in two different types of whole wheat flours used for the preparation of chapattis. In general, as the rate of flour extraction increases, the phytic acid content also increases. *Maida* (wheat flour of low extraction) on the other hand, contains much less phytic acid than WWF due to the removal of phytic acid rich parts of the wheat kernel during the milling process. The straight-run flour of the variety 'C 303' as tested by Austin and Ram (1971) contained only 26.36 mg per 100 g (dry matter) phytic phosphorus.

Phytic acid has both adverse physiological effects and health benefits in human nutrition. The adverse effects of phytic acid are attributed to its ability to form complexes with positively charged food components such as minerals and trace elements, proteins, and carbohydrates resulting in their reduced absorption or utilization and causing growth inhibition (Kumar et al, 2010). The interaction of phytic acid with minerals and other dietary nutrients is, however, pH dependent.

(a) ***Effect on mineral* absorption:** Phytic acid being a strong chelator binds the essential minerals such as calcium, magnesium, iron, zinc, copper, manganese, and phosphorus forming insoluble complexes thus adversely affecting their absorption and digestion. High concentration of calcium increases the antinutritive effect of phytic acid on the bioavailability of minerals and trace elements thus impeding the action of natural phytase.

(b) ***Effect on protein digestibility*:** Phytic acid forms strong insoluble complexes with some proteins which may affect the protein structure that can impede the enzymatic activity, protein solubility and digestibility. Phytic acid interacts with proteins at both acidic and alkaline pH. Phytate-protein complexes are more resistant to proteolytic digestion compared to protein alone at low pH and are thought to be responsible for the reduced bioavailability of the minerals (Lathia and Koch, 1989). The protein digestibility also depends upon the type of flour used for making chapattis, baking conditions such as use of leavening agent, pH of the dough, temperature and the rate of phytic acid hydrolysis by the phytase during the baking process.

(c) ***Effect on carbohydrate utilization*:** Phytic acid also forms complexes with carbohydrates of foods thereby reducing their solubility and adversely affecting the digestibility and absorption of glucose. Phytate- calcium ion complexes are also reported to inhibit amylase activity thus affecting the starch hydrolysis.

(d) ***Effect on certain vitamins*:** Phytic acid as an acid is capable to chelate with the vitamin niacin, the deficiency of which is known as pellagra (Anderson, 2005).

This chelating property of phytic acid with food components is particularly detrimental to consumers in developing countries who heavily rely on cereal grains and legumes to meet their nutrient needs. Nevertheless, the antinutrient impact of phytic acid may be reduced through soaking, germinating, fermentation, and cooking of cereal grains and legumes. During processing, the endogenous phytate- degrading enzyme causes dephosphorelation of the food thus facilitating the absorption of the nutrients. Chapattis are generally made from whole wheat flour (WWF) of near 100 per cent extraction. Phytic acid is reported to reduce significantly during chapatti-dough keeping (2 to 3 h) with additional reduction taking place during the chapatti-baking process (Swarnjeet et al, 1982; khan et al, 1986; Lathia and Koch, 1989; Syed et al, 1992). Agte and Joshi (1997) reported a 40 per cent decrease in Phytic acid when wheat chapatti dough was rested for 12 h and found significant decreasing trend with increasing resting time. Degradation of phytic acid due to chapatti making without resting of wheat chapatti dough was only 14 to 19 per cent. The extent of phytic acid degradation determines the extent of bioavailability of zinc and other minerals. Contrary to these reports, Mehdi and Abrol (1972) reported no destruction of phytic acid during the dough resting time and chapatti baking process. This anomaly may be due to the phytase status of the flour used by them. Swaranjeet et al (1982) have also shown that the phytic acid content of normal chapattis may further be reduced through the use of leavening agent and other supplements or by adjustment of dough pH. Since the amount of lactic acid used by them for adjusting pH was not detectable in the chapattis, its contribution to accelerate phytate hydrolysis is made more significant. The application of leavening agent and the use of supplements appear to induce endogenous enzymic hydrolysis of the phytates in chapattis. However, during baking of chapattis, the enzyme phytase gets deactivated. Notwithstanding, the balance amount of phytic acid left in chapattis, which upon ingestion may precipitate the above stated adverse effects thus limiting the bioaccessibility of essential nutrients.

Despite being labeled as culprit for the above stated adverse effects, phytic acid has potentially beneficial effects as well, such as:

- can act as antioxidant,
- exhibit some anticancer properties,
- has capacity to reduce cholesterol and triglycerides, and
- can positively impact the glycemic response of certain foods.

These positive effects of phytic acid may explain why high-fiber foods have cholesterol lowering effects. Further, these beneficial effects of phytic acid are evident from the fewer incidences of cancer, diabetes, renal problems, and arteriosclerosis in developing countries where people depend largely on plant-based diets especially cereals and legumes which contain a substantial amount of the phytate (Kumar et al, 2010).

***Effect of Dietary Fibre*:** It is now well known that the dietary fibre serves a very important role in human health in preventing cancer, cardiovascular diseases, diabetes, and several gastrointestinal disorders (Duss and Nyberg, 2004). On the contrary, the dietary fibre actually has several adverse antinutrient effects as well. It has adverse effects on bioavailability of minerals, protein digestibility, etc.

(a) ***Bioavailability of minerals*:** Dietary fibre impedes the bioavailability of minerals, most of which is attributed to the presence of phytic acid. The inhibitory effect of wheat bran in whole wheat flour used for making chapattis on iron absorption may be primarily attributed to its phytate rather than its dietary fibre. Reinhold et al (1981) have shown a direct relationship with persistent fecal loss of calcium, magnesium and zinc among Iranians with a very high consumption of unleavened whole wheat flour flat bread containing high fibre and phytate concentrations. However, in the presence of vitamin C and animal or fish protein in the diet, the impairment of iron and possibly other minerals absorption can be minimized (Harland, 1989). The vegetarians (children and lactating mothers in particular) who solely depend upon plants as source of protein, nevertheless, suffer the most from mineral malnutrition.

(b) ***Protein digestibility*:** The dietary fibre constituents may reduce protein digestibility and increase nitrogen excretion (Acton et al, 1982). The protein digestibility of chapatti and similar products is reported to be affected by the dietary fibre content of the wheat flour (Lathia and Koch, 1989).The extent of protein digestibility of flat breads and amino acids released depended upon the type of flour used, baking conditions, phytic acid content and other protease inhibitors in these breads. *Roghani naan* made from low extraction flour had protein digestibility of 87.6 per cent whereas chapatti made from whole wheat flour had a protein digestibility of only 80.0 per cent. A decrease in lysine content with increase in phytic acid concentration was observed in the case of chapatti and other breads after enzymic hydrolysis of their protein. Wheat bran fibre is reported to cause a significant loss of *in vitro* activity of lipase, trypsin and chymotrypsin (Schneeman, 1978).

***Effect of Lectins*:** Wheat like other cereals and legumes, contains lectins highly concentrated in wheat germ and are designated as wheat germ agglutinin (WGA). The level of WGA is reported to be between 100 to 500 μg/g in wheat germ, and between 29.5 to 50μg/g in whole meal flour (Punder and Pruimboom, 2013). The lectins are known to be associated with some detrimental physiological effects particularly in sensitive individuals. They can cause acute gastrointestinal symptoms, including nausea and vomiting. They can disrupt protein and carbohydrate absorption and metabolism. They are reported to contribute to the celiac disease, promote the growth of harmful bacteria, provoke numerous immune responses (allergies), and agglutinate erythrocytes leading to anemia (Thompson, 1993). Heat processing, however, can reduce the antinutrient effect of lectins since they are heat labile, but low temperature or insufficient cooking may not completely eliminate their adverse effects. Chapattis normally made from whole wheat flour may retain some activity of lectins since they are baked / cooked for a short period.

***Effect of Protease and Amylase Inhibitors*:** The protease inhibitors such as trypsin inhibitors inhibit the action of trypsin and other proteases in the gut preventing the digestion and subsequent absorption of protein ingested, and leading to protein deficiency and related health disorders. Bread wheat which is commonly used to make chapatti is reported to have twice as much trypsin activity as other wheat (Bora, 2014). Amylase inhibitors prevent the action of amylase enzymes that breakdown the starches and other carbohydrates, preventing the release of simple sugars and subsequent absorption by the body. Although, cooking may destroy or reduce the level of amylase inhibitors, even residual amounts of these may serve to reduce the rate of carbohydrate digestion and absorption. The enzyme inhibitors may also increase the amount of available carbohydrates that escape small intestinal digestion and enter the colon (Thompson, 1993).

***Effect of Polyphenolic Compounds*:** Wheat contains polyphenolics such as phenolic acids, flavonoids, condensed tannins and lignans, some of which at higher intakes are reported to be implicated in protection against several diseases. However, these compounds have the property to chelate minerals such as iron, zinc and reduce their absorption, and also inhibit the digestive enzymes leading to precipitation of proteins thus affecting their absorption (Thompson, 1994).

Improving Nutritional Quality of Chapatti

In view of the above factors affecting the nutritional quality of chapatti, following approaches are suggested to improve the nutritional quality of chapatti:

1. Plant breeding,
2. Modifications in processing methodologies,
3. Fortification/supplementation of whole wheat flour.

1. *Plant Breeding*: This agriculture approach must address issues related to micronutrient bioavailability to combat micronutrient malnutrition among consumers whose staple diets are predominantly cereal based. The objective should be to develop wheat varieties/cultivars that are rich in micronutrients (e.g. minerals) and substances that promote micronutrient bioavailability (e.g. ascorbic acid, S-containing amino acids, etc.), and decrease the level of antinutrients (e.g. phytates, polyphenolics, etc.) that inhibit micronutrient bioavailability (Welch and Graham, 2004). Austin and Ram (1971) tested wholemeal wheat flour (*atta*) from 17 wheat varieties for phytic phosphorus with values ranging between 105.55 ('C 303') to 146.60 ('NP 890') mg per 100g dry matter. This suggests the need for selecting wheat variety low in phytic acid for making nutritious chapattis. Research efforts should also be continued to raise the level of limiting essential amino acids in the new wheat varieties / cultivars under development.

2. Modifications in Processing Methodologies:

(a) ***Milling*:** The whole wheat flour preferred for chapatti making is generally milled in plate or stone mill which causes formation of higher levels of damaged starch resulting in the development of desirable typical wheaty aroma and sweetish taste of chapatti during baking. However, the high temperature developed during milling denatures the protein and decreases the essential amino acids and fatty acids content of flour and eventually in chapattis. Prabhasankar and Rao (2001) have shown that there are comparatively lower losses in essential amino acids and fatty acids in the case of whole wheat flour milled in plate mill as rise in flour temperature is comparatively lower than in that milled in stone mill. Milling of whole wheat flour for chapatti making should, therefore, be accomplished in plate mill. The whole wheat flour of different extractions is used to make chapattis. As the rate of flour extraction decreases phytic acid content also decreases as reported by Austin and Ram (1971). Sabir et al (1982) have also shown that removal of 10 per cent bran during milling resulted in 20 per cent reduction in phytic acid in two different types of whole wheat flours used for the preparation of chapattis. However, combined removal of 5 per cent bran and 10 per cent fines yielded atta which possessed better kneading, baking, puffing, and nutritional qualities than other combinations tested. Hatcher et al (1997), on the other hand, found 85 per cent extraction roller mill flour to yield chapattis with preferred sensory appeal. Removal of 20 or 40 per cent of the initial low ash flour streams resulted in blends that displayed improved objective textural properties and good to excellent sensory qualities. Therefore, the whole wheat flour of 85 to 90 per cent extraction should be recommended for making nutritious and excellent sensory quality chapattis. Several traditional household food-processing and preparation methods can be used to enhance the bioavailability of micronutrients in plant-based diets. These include thermal processing, mechanical processing, soaking, fermentation, and germination / malting.

(b) ***Baking:*** In order to reduce the adverse effects of phytic acid on the nutritive value of chapatti, it will be desirable to prepare dough in warm water and rest the dough for at least one hour (preferably 2 to 3 h) before baking. This will facilitate the phytase to hydrolyze the phytates to a greater extent. Swarnjeet et al (1982) achieved this by using leavening agent and lactic acid to reduce the pH of the dough. In normal chapatti dough having a pH of 6.1, phytate was reported to hydrolyze to the extent of 16 per cent only as compared to 62.5 per cent when the dough pH was adjusted to 5.1 because of the increased phytase activity. Also, the baking of chapatti should also be carried out under controlled temperature conditions (Yadav et al, 2008) to minimize losses in essential amino acids and vitamins particularly thiamin. In this respect, the thickness of dough sheet (raw chapatti) should also be optimized as said above.

3. Fortification/ Supplementation of Whole Wheat Flour: Wheat protein is incomplete qualitatively being limiting in essential amino acids especially lysine. If one or more amino acid is not present in sufficient amount, the protein in the diet is considered incomplete. Also, the micronutrients especially minerals in whole wheat flour are made inaccessible biologically due to the presence of antinutrients such as phytates and polyphenolics as stated above. Thus, heavy dependence on whole wheat flour for chapatti has resulted in the prevalence of micronutrient and protein-calorie malnutrition. By combining complementary proteins, the overall quality of the protein may be increased. Therefore, supplementing whole wheat flour with flours from legumes, oilseeds and other cereals have been recommended by several workers to combat the malnutrition problem. Because, these food materials are not only rich in essential amino acids but also provide ample amount of minerals. Fortification with suitable mineral substances, protein concentrates and isolates, and animal proteins has also been suggested. However, the successful development of supplemented flours for chapatti will depend upon the acceptance of the product. If not accepted, even the most nutritious food conceivable will be of no value to the malnourished. The properties that delineate the product acceptability may include the appearance, flavour/aroma and storage behavior of the supplemented flour and the dough, mixing and handling of dough, rolling and baking of the dough sheet, appearance/colour, aroma/flavour and texture of chapatti, etc.

***Supplementation with Other Cereals*:** The cereals other than *aestivum* wheat do not make chapattis of acceptable quality on their own. Blending cereals with one another may improve the acceptability of chapattis and the nutritional value.

Kaur and Hira (1988) have shown that supplementation of ***aestivum*** wheat with **durum** wheat in the ratio of 60:40 resulted in acceptable chapattis and an increase in crude protein, methionine, cysteine and available lysine content, and increase in trypsin inhibitor activity. When supplemented with 10 per cent Bengal gram flour or soya protein concentrate, there was further increase in the crude protein, available lysine, and trypsin inhibitor activity of the *aestivum*- durum blend.

Triticale rich in protein and lysine content was used to supplement whole wheat flour to improve the nutritional quality of chapattis. A blend of wheat flour, triticale flour and Bengal gram flour (40:40:20) exhibited higher values for PER, plasma protein and plasma lysine compared to other blends tested (Juneja et al, 1980). Chawla and Kapoor (1982) concluded that substituting wheat flour with more than 30 per cent triticale flour for chapatti making yielded chapattis having higher protein, lysine, feed efficiency ratio (FER) and protein efficiency ratio (PER) values. These substitutions may help in combating the protein malnutrition in the developing world.

Gujral et al (2012) prepared chapattis by incorporating germinated **brown rice flour** (10 per cent) into whole wheat flour. The flour blends and chapattis were evaluated for their antioxidant properties. Addition of germinated brown rice flour increased the total phenolic and flavonoid contents, and mineral chelating activity significantly. The antioxidant activity increased significantly by the addition of brown rice flour and addition of germinated rice flour further increased the antioxidant activity significantly. Baking of flour into chapattis caused significant decrease in phenolic and flavonoid contents. However, baking of the flour blends into chapatti increased the reducing power, metal chelating activity and antioxidant activity.

Finger millet or Ragi (*Eleusine coracana*, L.) is one of the most nutritious cereals being rich in essential amino acids, minerals, B-complex vitamins, dietary fibre, and host of complex bioactive polyphenolics having nutraceutical effects. *Ragi* has several health benefits like antioxidative, antidiabetic, anti-inflammatory, hypochlesterolaemic and antimicrobial (Devi et al, 2014). *Ragi* flour is gluten free, an advantage for gluten tolerant patients.Sharma et al (2017) blended finger millet flour into wheat flour in the ratio of 1:3 to prepare composite flour for nutritious chapattis. The composite flour had up to 30.7 per cent higher total phenolic content, 38.2 per cent higher total flavonoid content and 75.4 per cent higher antioxidant activity than the wheat flour. The slowly digestible starch correlated positively with total phenolic content and water absorption correlated positively with damaged starch content. Chapattis prepared from the composite flour showed lower retrogradation as evident from the mixolab retrogradation index, higher values of soluble starch and soluble amylose in stored chapatti. The chapattis made from the composite flour had higher slowly digestible starch and resistant starch values suggesting as potential vehicle for health promoting functional properties.

Thondre and Henry (2009) reported that **barley beta-glucan** significantly reduces the glycemic index of chapattis supplemented with doses of 4 to 8 g per serving. Sharma and Gujral (2014a) have shown that replacing whole wheat flour with whole **barley** flour (28 to 84 per cent) increased the total phenolic (TPC) and flavonoid (TFC) contents, and antioxidant activity of the blends. Baking of chapatti significantly decreased the TPC and TFC, however, antioxidant activity increased. The reducing power of the flour also increased significantly upon incorporation of barley flour. The polyphenol oxidase (PPO) activity also increased significantly in the blends; however, baking chapattis lowered the PPO. The mineral chelating activity was, on the other hand, found to increase. Narwal et al (2017) also observed significant increase in β-glucan content and antioxidant activity of chapatti at 30 per cent level of blending with **hulless barley flour**. Barley is high in protein, dietary fibre, a variety of minerals, and has the lowest glycemic index among the grains.

From these studies, it can be concluded that the antioxidant activity of chapattis can be enhanced, and that chapatti can be an ideal vehicle to deliver bioactive components of brown rice, finger millet and barley.

Amaranth (*Amaranthus paniculatus*) grain, a pseudo-cereal contains a substantial quantity of protein and fat, and is high in lysine (5.19g/16g N) as compared to most cereals (Singhal and Kulkarni, 1988).It is a good source of dietary fibre, and minerals such as iron, magnesium, phosphorus, copper and manganese.The amaranth grains also contain many bioactive compounds such as phytosterols, polyphenols, etc. with health promoting properties including decreasing plasma cholesterol, stimulating immune system, antitumor activity, reducing blood sugar, etc. (Chauhan et al, 2016). In order to improve the quality and quantity of protein in wheat flour chapatti, Sarojini et al (1996 b) supplemented wheat flour with amaranth flour in the ratio of 3:1. The blend contained 12.3 per cent protein, 2.34 per cent fat and 12.55 per cent dietary fibre. Chapattis scored high for appearance, taste, flavour, texture and overall acceptability. According to Banerji et al (2017), the chapattis prepared from 40 per cent amaranth flour blend with wheat flour were significantly rich in iron, calcium, manganese, fat, protein, and lysine with improved *in vitro* digestibility.

Tartary Buckwheat (*Fagopyrum tartricum*) protein has a well-balanced composition of amino acids especially threonine, lysine and tryptophan. It is a very good source of polyunsaturated fatty acids, dietary fibre, and minerals such as calcium, magnesium, potassium, iron, manganese, zinc, copper and selenium. It is an excellent source of the antioxidants rutin and quercetin (Fabjan et al, 2003). Chapattis supplemented with Tartary buckwheat flour at the acceptable level contained increased amounts of crude protein, crude fat, total ash, crude fibre and minerals like iron, potassium, phosphorus and selenium. Chapattis were enriched with the antioxidants especially rutin and quercetin commonly found in fruits (Bano et al, 2014).

Non-wheat multi-wholegrains (barley, maize, sorghum, pearl millet, finger millet, soybean, Bengal gram, horse gram, and amaranth), which are rich in the nutritional components in which wheat is deficient, were used to develop multi-wholegrains supplemented *atta* (Pande et al, 2017). The chapattis made from composite flour containing 10 per cent multi-wholegrains were found sensorially acceptable. The addition of multi-wholegrains enhanced the nutritional quality of chapattis in terms of protein, minerals, dietary fibre, and amino acids contents.

***Supplementation with Legume / Pulse Flours*:** Legumes are rich in protein, minerals, dietary fibre and complex carbohydrates having low glycemic index. Legumes protein is high in essential amino acids, e.g. lysine but is deficient in sulphur-containing amino acids making legumes a potential complement to cereals, e.g. wheat. However, the presence of antinutritional factors such as, trypsin inhibitors, oligosaccharides, lectins, and tannins limit the availability of nutrients present in them. Nevertheless, processing like dehulling, soaking, cooking, etc. are reported to improve the nutritional properties. Soybean and many other pulses, e.g. chickpea, cowpea, pigeon pea, black gram, *rajmah* (kidney bean), rice bean, *khesari*, fenugreek seeds have been variously used to supplement whole wheat flour for making chapattis in order to step up the nutritive value of chapattis.

Defatted soy flour has generally been used to fortify whole wheat flour to nutritionally enrich chapattis. Defatted soy flour is rich in protein and available lysine, ash, crude fibre, calcium, phosphorus and iron. Supplementation with defatted and heat treated soy flour has been reported to enhance the protein, calcium, phosphorus, and iron contents, and PER and in vitro digestibility of chapattis (Bhat and Vivian, 1980; Lindell and Walker, 1984; Rawat et al, 1994). However, soy fortified chapattis retained some trypsin inhibitor activity. Wheat *roti* (unleavened Pakistani bread) supplemented with soybean flour had increased level of protein, fat, ash, calcium, iron, phosphorus and phytic acid contents compared to the control. The carbohydrate content of *rotis* significantly decreased whereas crude fibre content remained the same. Protein efficiency ratio, net protein utilization and biological value of supplemented *rotis* were higher than those of the control; however there was no significant effect of supplementation on the true digestibility of wheat *rotis*. Addition of soybean flour up to 20 per cent level did not cause any adverse effects on the sensory acceptability of *rotis* (Maqbool et al, 1987). Khetarpaul and Goyal (2008) reported improvement in the in vitro protein and starch digestibility of soy fortified chapattis after giving processing treatment (mixing and roasting) to the soy-wheat flour blends. Khan et al (2009) conducted rat bioassay to evaluate the nutritional quality of chapattis prepared from wheat and soy flour blends. Blending with heat treated soy flour (10 per cent) significantly increased the crude protein, ash, crude fibre, and fat contents of the wheat flour. Incorporation of soy flour to wheat flour significantly improved the minerals (iron, zinc, copper, calcium, magnesium, phosphorus and potassium) and amino acids (threonine, isoleucine, leucine, lysine, phenylalanine, valine, histidine and alanine) profiles, and the blends showed higher feed intake, food efficiency ratio, and biological value. These studies inferred that substitution of wheat flour with heat treated soy flour (both full fat and defatted) up to 10 per cent produced acceptable and nutritionally superior chapattis than the traditional wheat flour chapattis. Kadam et al (2012) used full fat soy flour prepared from soaked and autoclaved soybean grains and reported increase in protein quality and quantity, and mineral content of soy fortified chapattis. There was further increase in the nutritional quality of chapattis in terms of minerals (calcium and iron in particular) and dietary fibre when the wheat-soy blend was supplemented with 5 per cent dried fenugreek leaves powder as well.

Chickpea or Bengal gram (*Cicer arietinum*) has been traditionally used to supplement wheat chapattis. Supplementation with chickpea flour at 10, 15 and 20 per cent level improved the protein quality and quantity of *rotis* without affecting significantly their organoleptic qualities (Haq and Chaudhry, 1976). Supplementation of legumes to cereals improved calcium utilization as shown by Gupta and Kawatra (1992) in chapattis prepared from wheat: chickpea (80:20 and 70:30) combinations. Chapattis supplemented with chickpea flour have also been reported to have significant improvement in lysine and minerals (calcium, magnesium, iron, copper, and manganese) content, and biological value, net protein utilization, net protein retention and protein efficiency ratio (Sharma et al, 1995; Siddique et al, 1996).

Cowpea (*Vigna sinensis*) or *lobia* is rich in protein, fibre, lysine, minerals (potassium, calcium, magnesium, phosphorus, zinc, copper, manganese and selenium), and vitamins (A, C, several B-group vitamins). The high lysine content (486 mg / g nitrogen) makes cowpea an excellent enhancer of protein quality when combined with cereal grain proteins which are low in lysine but rich in sulfur amino acids (Prinyawiwatkul et al, 1996). However, it has not been fully utilized especially in developing countries because of the presence of antinutritional factors such as trypsin inhibitors, flatulence causing oligosaccharides, etc. Cowpea flour may be used to supplement chapattis after suitable treatments to lower the effect of antinutritional factors and the beany flavour (Okaka et al, 1979). Sharma et al (1999a) reported that supplementation with cowpea flour improves the protein, lysine, minerals and vitamin contents of chapattis.

Pigeon pea (*Cajanus cajan*) commonly known as *toor* or *arhar* is rich in protein, essential amino acids, iron, and folate. Chapattis prepared from 5 to 15 per cent supplemented flours were acceptable. The nutrient content particularly protein, iron and phosphorus increased with increased substitution (Harinder et al, 1999).The protein efficiency ratio of supplemented flat bread was significantly higher than that of the control (Sharma et al, 1995).

Ricebean (*Vigna umbellate*) is rich in protein, available lysine, calcium and zinc. Kaur and Mehta (1992) prepared acceptable chapattis by blending upto 10 per cent ricebean flour with whole wheat flour. Results revealed increase in protein, ash, crude fibre, and minerals (calcium, iron, zinc, and copper) contents. There was a significant increase in available lysine, a limiting amino acid in wheat.

Khesari (*Lathyrus sativus* L) known as Indian vetch, is valued as a nutritious legume primarily owing to its high protein content (26.5 to 28.7 per cent) and especially as a good source of some of the essential amino acids (Kay, 1979). Although, it is tasty to eat besides being nutritious, it has a drawback because of the presence of a neurotoxin, which can, however, be removed prior to use by steeping in hot water and repeated rinsing (Rehman et al, 2006b). The composite flour containing up to 20 per cent treated khesari flour yielded acceptable chapattis. The biological value, net protein utilization, digestibility, net protein efficiency ratio of chapattis increased as a result of improvement in protein quality and depletion of the toxin.

Black gram (*Vigna mungo)* is very nutritious as it contains high levels of protein, fibre, and minerals like calcium, iron, potassium, phosphorus, zinc, and vitamins like niacin, thiamin, riboflavin, folic acid. It is commonly used as pulse in several Indian cuisines. It is beneficial in mitigating elevated cholesterol levels, and cardiovascular problems. The neutral detergent fibre isolated from black gram is reported to possess significant hypolipidemic, hypoglycemic, and anticolon-cancer activity (Indira and Kurup, 2003). Wani et al (2016) supplemented the wheat chapatti with black gram flour in order to improve its quality and quantity of protein, and enhance the minerals and vitamins contents. Chapattis remained acceptable up to 15 per cent addition of black gram flour.

Fenugreek (*Trigonella foenum-graecum* L.) leaves (both green and dried) and seeds form an important ingredient in Indian cooking. Fenugreek seeds are rich in proteins, available lysine, and mucilaginous and total fibre. Seeds have medicinal properties especially being hypocholesterolemic and hypoglycemic. Darak Priya and Pawar (2015) prepared chapattis with partial replacement of wheat flour with fenugreek malt (seeds germinated, dried and powdered). Chapattis prepared from the blend containing 15 per cent fenugreek malt received the highest sensory scores. There was significant increase in protein and ash content and decrease in carbohydrate content of chapattis. The fenugreek malt was also found to reduce total serum cholesterol, total lipids and LDL cholesterol. Robert et al (2015) supplemented refined wheat flour with fenugreek seed powder (10 per cent) and reported significant reduction in the glycemic index and glycemic response of flat breads.

***Supplementation with Oilseed Flours/Meals*:** The use of protein rich full fat or defatted oilseed meals/flours show promise in improving the nutritive value of the supplemented products, at the same time optimum utilization of these flours. Oilseed flours/meals are reported to significantly improve the quality of wheat flour because of their high contents of protein, essential fatty acids and essential amino acids especially lysine (Bhat and Vivian, 1980). Several attempts have been made to enrich chapatti using oilseed flours without sacrificing its functional and sensory characteristics.

Bhat and Vivian (1980) prepared acceptable chapattis by incorporating defatted **peanut flour** and reported increase in protein and available lysine contents as the addition of peanut flour was increased. Khalil and Chughtai (1982) reported increased contents of protein, fibre, ash, and minerals in the *rotis* fortified with **peanut flour**. Lysine was still the limiting amino acid but there was increase in chemical score with 30 per cent substitution. The nutritive value, as judged by gain in body weight of rats, PER, NPU and biological value, increased significantly over the control *roti* at the 20 per cent level which was the optimum level of substitution.

Cottonseed flour is rich in protein with well-balanced amino acid profile. However, chapattis fortified with defatted cottonseed flour were not found acceptable because of the unpleasant dark colour, cottonseed flour aroma, and leathery texture suggesting further refining of cottonseed flour to extract the undesirable colour and flavour. Nevertheless, there was increase in protein and available lysine contents of chapattis (Bass and Caul, 1972; Bhat and Vivian, 1980). Rasool et al (2005), on the other hand, reported increase in protein digestibility, net protein utilization, protein efficiency ratio, net protein retention, and biological value when chapattis fortified with cotton seed flour (10 per cent) were fed to albino rats. Feed efficiency was also found to improve significantly. The cottonseed flour used by them was defatted, detoxified, low in fibre, and the fortified chapattis were acceptable to the taste panel.

Sunflower seed meal is a palatable source of high quality protein and fiber. The kernels are an excellent source of vitamin E and phytosterols, very good source of selenium, vitamin B1, and a good source of magnesium, potassium, manganese, copper, phosphorus, vitamin B6, folate, niacin, phytochemicals, as well as essential fatty acids (Holiday and Phillips, 2001). In order to enhance the nutritional quality of wheat chapattis, the whole wheat flour was supplemented with10 per cent defatted sunflower seed flour to prepare acceptable chapattis. (Rasool et al, 2005; Bhise and Kaur, 2013). Rasool et al (2005) detoxified and lowered the fibre content of sunflower seed flour as well. They fed the supplemented chapatti diet to albino rats to carry out the biological assay, and reported increase in protein digestibility, net protein utilization, protein efficiency ratio, net protein retention, and biological value with significant improvement in feed efficiency. Bhise and Kaur, on the other hand, used the texturized form of the defatted sunflower seed flour and found increase in the contents of protein, fat, fibre and ash, and rise in protein digestibility of chapattis as compared to those prepared from whole wheat flour.

Mustard seeds are rich in protein, fat, dietary fibre, phytonutrients, and are excellent source of selenium, very good source of manganese and omega-3 fatty acids, and a good source of magnesium, phosphorus, copper, iron, potassium, zinc and B-group vitamins. Rasool et al (2005) used the defatted, detoxified and low in fibre mustard seed flour to supplement whole wheat flour at 10 per cent level to prepare acceptable chapattis. The fortified chapatti diet when fed to albino rats was found to increase the biological assay values namely protein digestibility, net protein utilization, net protein retention, protein efficiency ratio, biological value with significant improvement in feed efficiency.

Flaxseed or linseed (*Linum usitatissimum*) is a pleasant tasting, slightly nutty and sweet whole grain. It is rich in protein, fat, dietary fibre, ash, several neutraceutical components including omega-3 fatty acid, and is well documented for its several health benefits (Wu et al, 2007).The nutritive value of flaxseed is shown in **Table 5**. In view of its pleasant taste and high nutritive value attempts have been made to prepare supplemented wheat based baked products. Hussain et al (2012) prepared unleavened flat breads with acceptable level of flaxseed flours (12 and 16 per cent full fat and partially defatted, respectively). These additions led to increase in total dietary fibre, and essential fatty acids and amino acids especially lysine in fortified flat breads. True digestibility (TD), net protein utilization (NPU), protein efficiency ratio (PER) and biological value (BV) improved significantly when diets containing flaxseed supplemented flat breads were fed to rats. The diet prepared from flat bread containing 16 per cent partially defatted flaxseed flour was biologically superior among other experimental diets due to higher values for TD, NPU, PER and BV. Studies conducted by Bhise and Kaur (2013) showed defatted flaxseed meal to contain 38.24, 12.24, and 2.71 per cent protein, fibre, and fat respectively, and *in vitro* protein digestibility of 43.77 per cent.

Thus, addition of texturezed defatted flaxseed flour in chapattis resulted in significant improvement in crude protein, crude fibre, ash and *in vitro* protein digestibility.

Table 5: Nutritive value of flaxseed (per 100g)

Proximate	Value (g)	Mineral	Value (mg)	Vitamin	Value (mg)
Protein	18.3	Calcium	255	Thiamin	1.7
Fat	42.2	Magnesium	392	Riboflavin	0.2
Carbohydrate	28.9	Phosphorus	642	Niacin	3.1
Dietary fibre	27.3	Potassium	813	Pantothenic acid	0.9
Ash	3.7	Iron	5.8	Folate	87 µg
Energy	534 kcal	Zinc	4.4		

Source: USDA (2006).

***Supplementation with Fruits and their Products*:** Dehydrated fruits and their products that are rich in protein and minerals have been used to supplement wheat chapattis to upgrade their nutritive value.

Jamun (*Syzygiumcumini*, L) also known as Indian blackberry is rich in minerals, vitamins, phytochemicals like flavonoids and anthocyanins, hence designated as neutraceutical. It has several health benefits and possesses antioxidant, antidiabetic and anticancer properties (Chaudhary and Mukhopadhyay, 2012). Kapoor et al (2015) supplemented wheat chapattis with hot-air-dried or freeze dried *jamun* pulp powder at maximum (3 per cent) acceptable level for improving nutritional quality. *Jamun* powder incorporation in chapattis slightly improved the fibre and ash contents. Freeze-dried *jamun* powder supplemented chapattis showed greater increase in bioactive compounds like anthocyanins, total phenols, and the antioxidant activity. These observations inferred that *jamun* fruit powder can be used as a potential source to improve the antioxidant status of chapattis.

Melon (*Cucumis melo*) seeds are a rich source of protein, oil, and minerals like calcium, phosphorus and iron (Teotia and Ramakrishna, 1984). Kaur et al (1988) prepared fortified chapattis with most acceptable combination of wheat flour and melon seed flour (80:20), which when fed to weanling rats exhibited increase in protein efficiency ratio as compared to the control but the differences were not significant. Supplementation with melon seed flour resulted in increase in protein and fat content of chapattis by 4.46 and 9.98 per cent respectively thus improving the energy content. However, lysine remained the limiting amino acid.

Jackfruit (*Artocarpus heterophyllus* L.) seeds are rich in protein, starch, minerals, B group of vitamins and phenolic compounds. This seed flour is gluten-free and can serve as a substitute for wheat flour for those with specific food allergies. Jackfruit seed flour is rich in nutrients as shown in **Table 6**.

Table 6: Nutrient composition of jackfruit seed flour (per 100g dry matter basis)

Parameter		Value
Protein	g	13.50
Fat	g	1,27
Ash	g	2.70
Fibre	g	3.19
Carbohydrate	g	79.34
Sodium	mg	6.0
Potassium	mg	1478
Calcium	mg	308
Iron	mg	13.0
Copper	mg	1.0
Calorific value	kcal/100g	382.79

Source: Ocloo et al (2010).

Sultana et al (2014) prepared chapattis from composite flours made of whole wheat flour + white jackfruit seed (lye-peeled) flour + brown jackfruit seed flour + Bengal gram flour in the proportions of 80:5:10:5, 55:15;15;15 and 45:25:20:10 and the control which were fortified with same amount of fat, salt, sugar and baking powder. It was concluded that chapatti containing 80 per cent whole wheat flour, 5 per cent white jackfruit seed flour, 10 per cent brown jackfruit seed flour plus 5 per cent Bengal gram flour was the best among the three blends because of its higher calorific value and overall sensory acceptability. The protein, ash, crude fibre and fat content of chapattis increased with increased level of jackfruit seed flour.

Fruit peels are good sources of dietary fibre, polyphenols, carotenoids and other bioactive compounds which have various beneficial effects on human health. **Banana peel** is rich in dietary fibre, proteins, minerals especially calcium, potassium and iron, and possesses antioxidant properties owing to the presence of polyphenolic compounds (anthocyanins, flavonoids, etc.) and carotenoids (Nagarajaiah and Prakash, 2011). Banana peel powder (BPP) prepared by Kurade et al (2016) was found to be a good source of protein, and an excellent source of dietary fibre and minerals. However, the principle component of BPP was carbohydrate. When supplemented to whole wheat flour, there was no significant difference in overall acceptability of chapattis up to 10 per cent addition of BPP. The total phenolic and flavonoid contents and antioxidant activity increased with increase in the concentration of BPP in chapattis.

***Supplementation with Green leafy Vegetables*:** Green leafy vegetables are micronutrient dense nature's gift to mankind. They are a good source of high quality protein, vitamins, and minerals especially calcium and iron. They also contain an immense variety of bioactive, non-nutritive health enhancing factors such as antioxidants, phytochemicals, essential fatty acids and dietary fiber.

Sawhney and Kawatra (1986) improved the protein quality of wheat chapatti by supplementing with **beet** (*Beta vulgaris*), **knol-khol** or *ganth gobhi* (*Brassica oleracea*) and **turnip** (*Brassica rapa*) green leaves in the form of *sag* – a cooked culinary pasty preparation containing cooked green leaves paste, shallow fried in ghee along with condiments and spices. The dehydrated *sags* prepared from beet, knol-khol, and turnip leaves respectively, contained 15.6, 21.3, 20.4 per cent crude protein, and 12.8, 11.6, 10.8 per cent ash, 297.9, 459.7, 497.8 mg/100 g ascorbic acid and 4276, 3204, 6119 μg/100g β-carotene (on dry matter basis). The lysine and tryptophan contents of beet, knol-khol and turnip sags were higher as compared with wheat chapatti. The dehydrated whole wheat flour chapatti, and different *sags* were blended in 4:1 ratio and fed to rats after fortifying with required amount of minerals and vitamins. Feed intake, body weight gain, protein efficiency ratio and plasma protein concentration increased with the supplementation of wheat chapattis.

Luthra and Sadana (1995) similarly prepared the *sags* using **bathu** (*Chenopodium album*), **carrot** (*Daucus carota*), **radish** (*Raphanus sativus*) and **spinach** (*Spinacia oleracea*) green leaves for supplementing wheat chapatti to improve its protein quality. The dehydrated powders of whole wheat flour chapatti and different *sags* were blended in 4:1 ratio and fed to rats after fortification with required amount of vitamins and minerals, etc. for biological assay. Supplementation of wheat chapatti with different *sags* improved the PER, plasma protein concentration and apparent protein digestibility, maximum being in the case of *bathu sag*. The chemical score of wheat chapatti improved for lysine and tryptophan contents with the addition of these *sags*.

Punia et al (2004) used freshly chopped **amaranth** (*cholai*) leaves in the ratio of 2:3 and 1:2 to enhance the nutritive value of the flat breads *paratha* and *poori*, respectively. The nutrient composition of amaranth leaves and the supplemented products are given in **Table 7**. The difference in nutrient content between the two products may be due to more of fat absorbed by poori during deep fat frying.

Table 7: Nutrient composition of amaranth and supplemented *paratha* and *poori* (Per 100g fresh weight basis).

Property /Product	Fresh leaves	*Paratha*	*Poori*
Moisture g	80.48	25.30	9.93
Protein g	5.44	12.26	10.34
Fat g	0.34	5.42	25.91
Minerals g	3.96	2.98	2.38
Crude fibre g	1.56	1.22	0.23
Carbohydrate g	8.22	52.83	51.44
Calcium mg	611.95	177.73	114.84
Iron mg	1.75	4.72	1.35
Ascorbic acid mg	33.26	5.10	4.87
B-carotene μg	2744.00	1592.00	1540.50

Source: Punia et al (2004.

Shanthala and Prakash (2005) prepared chapattis by incorporating 5 or 10 per cent dried **curry leaf** powder (CLP) into wheat flour. The addition of CLP affected the colour and appearance of chapattis. However, at 5 per cent level of incorporation, the texture, odor and taste of chapattis were not affected. Since curry leaf powder is rich in protein, fibre, iron, phosphorus and calcium, the nutritive value of chapattis improved accordingly.

Joshi and Mathur (2015) prepared leaf powder mixtures comprising dehydrated leaves of **beet root, carrot**, **cauliflower** (*Brassica oleracea*) and **turnip** in 1:2:1:1 ratio, respectively. Leaf powder mixture (LPM) was analyzed for proximate composition, minerals (calcium, phosphorus, iron, copper, zinc, manganese, magnesium) and antinutritional factors (oxalates and phenols). LMP contained appreciable amount of protein, fibre, carbohydrates, and minerals with generally low levels of antinutients. Chapattis supplemented with LPM were assessed for quality on the basis of sensory attributes and were well accepted up to 10 per cent addition of the LPM. Protein, iron and calcium contents were significantly higher in LPM substituted *missi roti* or chapatti.

Spinach (*Spinacia oleracea*) is a very nutrient-dense vegetable. It is an excellent source of vitamins, minerals especially iron, antioxidants and other phytonutrients such as carotenoids (β-carotene, lutein and zeaxanthin) and flavonoids. The spinach chapattis prepared by incorporating fresh blanched spinach paste showed the presence of higher levels of protein, fat and minerals especially iron and zinc as compared to the control. Chapattis also got enriched with carotenoids, chlorophyll and vitamin C derived from spinach (Khan et al, 2013). Khan et al (2015) also developed chapatti premixes by blending whole wheat flour with spinach powder prepared by blanching and dehydrating spinach leaves. Incorporation of 5 per cent spinach powder to wheat flour was found to be optimum for chapatti making. Addition of spinach powder significantly increased total carotenoids, total chlorophyll, vitamin C, iron, calcium and magnesium contents in the premixes and chapattis. However, there was significant decrease in total carotenoids, total chlorophyll as well as vitamin C contents during baking of chapattis. Although the level of dehydrated spinach powder incorporation is small, the increase in the nutrient content is remarkable. Thus, dehydrated spinach powder chapattis have the potential to serve as the valuable source of calcium, iron, fibre, vitamin C and carotenoids in the diet of the population in the Indian subcontinent.

***Supplementation with Herbs*:**

Aloe vera (*Aloe barbadansis,* Miller) is well known for its medicinal and therapeutic properties due to the presence of bioactive components in the gel of the leaves. The Food and Drug Administration of USA refers aloe vera as a food, herb or dietary supplement. In the food industry aloe vera is being exploited as a functional ingredient in the preparation of health drinks and beverages. Bhargawa et al (2014) prepared chapattis from whole wheat flour supplemented with 3 per cent aloe juice for maximum acceptability. Aloe juice raised the level of minerals, ascorbic acid and total phenolic content of the chapattis.

***Supplementation with Protein Concentrates and Isolates*:**

Fish protein concentrate (FPC) is a valuable supplement to plant protein based diets. Relatively small quantity of good quality, odorless fish protein has a great potential to supplement particularly the cereal based diets because these are deficient in essential amino acids especially lysine. Munro et al (1969) have shown that addition of 5 per cent FPC resulted in increase in protein, and total and available lysine and total methionine contents in wheat chapattis. There was also a significant increase in feed intake, body weight gain, and protein efficiency ratio when FPC fortified chapatti diet was fed to rats. Chari and Sreenivasan (1980) prepared FPC from shark meat which contained 87.6 per cent protein with balanced essential amino acid composition. Fortification with 2 per cent FPC yielded chapattis of acceptable quality with consequent increase in essential amino acids.

Casein co-precipitate (CCP) prepared from milk has capacity to immobilize water by hydration, and can be used to improve the texture and protein quality of food products. Chapatti fortified with optimum acceptable level (20 per cent) of CCP was mixed with the experimental diet for biological assay. There was increase in feed intake, gain in body weight, protein efficiency ratio, biological value and digestibility coefficient when the experimental diet contained CCP incorporated chapattis in comparison to CCP free chapatti diet (Vijayalakshmi et al (1983).

***Supplementation with Dairy By-products*:** Supplementing unleavened flat bread with **whey protein concentrate** powder and **butter milk** powder resulted in increase in protein, ash and minerals (calcium, potassium, magnesium and phosphorus) contents (Madenci and Bilgicli, 2014). Srivastava et al (2015) reported improvement in the nutritional value of chapattis in terms of enhanced protein, fat and ash contents when liquid butter milk was incorporated into chapatti dough. Butter milk is a good source of proteins, carbohydrates, minerals, and vitamin-A, tocopherols and phospholipids.

Fortification with Minerals

The extent to which a mineral is absorbed from the diet depends upon a number of dietary and physiological factors such as amount ingested; the biological effectiveness of the form of the element ingested, and the presence of other factors in the diet which may inhibit or promote the availability of that mineral (FAO/WHO, 2004). It is widely recognized that minerals are generally less bioavailable from foods of plant origin than from animal foods. Grover et al (1994) have shown the in-vitro availability of iron, copper and zinc in chapattis to the extent of about 47, 69 and 93 per cent, respectively. The phytic acid, dietary fibre and polyphenolics found in cereals are thought to be the major contributors to reduced bioavailability of minerals from food derived from cereals. Both the level and source of dietary protein are also reported to be important determinants of iron availability. Therefore, it is desirable to fortify cereal based foods with relevant mineral to make up the deficiency. Fortification of cereal flours with minerals requires that the mineral form:

1. should not alter the processing and storage quality of the flour, and sensory quality of the food product,
2. should have high bioavailability value, in other words should be adequately absorbed,
3. should be low in the inhibitory effect of dietary components such as phytic acid, phenolic compounds, and calcium (Hurrel, 2002).

***Fortification with* Iron:** The iron compounds commonly used include elemental iron, ferrous sulphate, ferrous fumarate, ferric sodium ethylenediaminetetraacetate (NaFeEDTA), etc. However, there is little direct evidence that elemental iron powders have a beneficial effect on the status of iron. Ferrous sulphate though soluble in water and cheaper to use but has negative property of provoking fat oxidation during storage. The better absorbed alternative iron compounds for cereal flour fortification include encapsulated ferrous sulphate and NaFeEDTA which do not provoke fat oxidation during storage of cereal flours and products prepared there from.

Elwood et al (1970) studied the availability from chapatti of **wheat iron**, and of **iron of ferrous chloride** added to wheat flour, given to consumers as part of their usual diet. The wheat iron was found to be better absorbed than the iron derived from ferrous chloride but the bioavailability of both was very low. The mean absorptions of wheat iron were 4.0 and 2.2 per cent from white flour and whole meal chapatti, respectively, and 2.1 and 1.8 per cent from the iron salt baked into similar chapatti. They suggested the use of an iron preparation that can be added to flour in relatively high concentrations without seriously affecting its keeping or baking properties. Prabhavathi and Narasinga Rao (1979) using the *in vitro* method reported that the **ionisable iron** from chapatti made from refined wheat flour was almost three times more available than from that made from whole wheat flour. Murthy et al (1985) reported that though chapatti contained 4.9mg/100g total iron, its absolute availability without fortification was only 0.35 per cent, which marginally increased to 0.39 per cent with the use of **iron fortified salt** (iron content 0.2 mg). Addition of tannin reduced the absolute availability by 20 per cent. Grewal et al (2000) investigated the availability of **iron** from differently processed and cooked wheat products using hemoglobin regeneration efficiency (HRE) method. The protein, fat, fibre, phytate and total phenolics of pearled wheat flour (PWF) were lower than whole wheat flour (WWF). All minerals analyzed except zinc were significantly higher in WWF while soluble iron was higher in PWF. Baking flours into chapatti resulted in significant increase in HRE, and HRE of chapatti was higher than that of *paratha*. There was significant correlation between soluble iron and HRE.

The absorption of iron is dependent on its interaction with other dietary components which either inhibit or stimulate it. Christian and Seshadri (1989) studied the effect of addition of **tea, milk and ascorbic acid** to a *bhakri* (wheat flour chapatti) meal for *in vitro* availability of iron. Tea added to the meal had a significant depressing effect on available iron. Addition of ascorbic acid or milk to the meal completely counteracted the inhibitory effect of tea, and addition of both ascorbic acid and milk brought about an even greater enhancement in the *in vitro* availability of iron than when either was added alone. The consumption of tea along with cereal meals, particularly at breakfast is common practice in the Indian subcontinent. Tea being rich in polyphenolics and tannins is potent in inhibiting the absorption of iron. Simultaneous consumption of ascorbic acid containing foods is reported to counteract the negative effect of tea. Similarly, milk has been shown to promote iron availability *in vitro* from cereal meals.

Kloots et al (2004) evaluated the relative availability of different sources of **iron** from iron fortified whole-grain wheat flour which was baked into chapattis to determine the *in vitro* dialysability and absorption. Relative to iron from ferrous sulphate, iron from NaFeEDTA was found to be 7 times more available in the *in vitro* assay. Iron absorption from digested chapattis fortified with NaFeEDTA was significantly higher than from digested chapattis fortified with ferrous sulphate. The authors concluded that only the addition of NaFeEDTA to whole-grain wheat flour may result in a food product that may be expected to effectively improve the iron status of people consuming it.

Wijngaart and Codling (2013) under Flour Fortification Initiative have presented a report "**impact of flour fortification on common Asian wheat flour-based foods**" as per the technical guidelines of the World Health Organization (2009). The chapattis made with iron fortified *atta* were slightly darker in colour than the control chapatti. Chapatti made with flour fortified with ferrous sulphate was denser than chapatti made with flour fortified with NaFeEDTA and electrolytic iron or control. Chapattis made with ferrous sulphate fortified flour retained the most moisture, followed by chapatti made with the control flour and those made with flours fortified with NaFeEDTA and electrolytic iron. Chapattis made with flour fortified with NaFeEDTA was the most preferred in overall rank.

***Availability of Zinc*:** Wheat contains 2.24mg/100g, whereas chapatti made from it has 2.21mg/100g zinc. The availability of zinc from wheat products, e.g. chapatti is poor as indicated by high phytate/zinc molar (32mM/100g), and phytate x calcium/ zinc molar (37mM/100g) ratios. Wheat-legume combination (e.g. chapatti-lentil 5:1) is a better source of available zinc than wheat alone as indicated by lower phytate/zinc molar (14mM/100g) and phytate x calcium/zinc molar (18mM/100g) ratios (Hira and Kaur, 1993). Further studies were carried out to test the wheat flours (whole wheat and pearled wheat) and cooked products made from them for phytate: zinc, phytate x calcium: zinc, phytate: iron and iron: zinc ratios to evaluate mineral availability (Grewal et al, 1999). Results showed that zinc bioavalability could be maximum from whole wheat flour followed by pearled wheat flour as per phytate: zinc and phytate x calcium: zinc ratios obtained from these flours. In spite of high phytate content in chapattis, the phytate: zinc, phytate x calcium: zinc, phytate: iron and iron: zinc ratios were found to be lowest in chapattis due to proportionately higher mineral content. These low ratios indicated better availability of zinc and iron from chapattis made from both whole and pearled wheat flours. Chapattis made from pearled wheat flour (PWF) and whole wheat flour (WWF) were tested for bioavailability of zinc using rat bioassay (Grewal and Hira, 2003). Results revealed higher zinc absorption/retention from chapatti diets compared to the respective flours and *paratha* diets. However, percent zinc absorption and femur zinc contents of rats fed on PWF chapatti diet were significantly higher compared with WWF chapatti diet. Based on femur weight, femur zinc, plasma zinc and zinc absorption and retention, it was concluded that pearled wheat diets were better than the whole wheat diets, and among the products, chapatti was better carrier than *paratha*.

Ahmed et al (2008).fortified whole wheat flour with calcium carbonate, ferrous sulphate and zinc sulphate at suitable levels and the chapattis prepared from the fortified flours were subjected to rat bioassay to determine the bioavailability of calcium, iron and zinc. The absorption was higher in the rats fed on fortified chapatti diet than those fed on unfortified chapatti diet. The minerals were found to interact with each other resulting in decrease of bioavailability among each other. However, calcium absorption was higher when chapattis were fortified with calcium plus zinc than when chapattis were fortified with calcium plus iron. Nevertheless, it is feasible to fortify whole wheat flour with suitable combinations of calcium, iron or zinc to combat malnutrition due to the deficiency of these minerals.

Bhavyashree et al (2009) examined composite meal comprising wheat chapatti, potato, French beans and condiments/spices for i*n vitro* bioaccessibility of iron, zinc, calcium, copper and manganese. The meal on fresh weight basis contained total iron 2.31, zinc 0.83, calcium 8.6, copper 0.22 and manganese 0.30mg/100g of which only 1.6, 4.7, 5.1, 16.6 and 24.5 per cent of the total, respectively was bioaccessible. Hemalatha et al (2007b) have shown that pressure cooking and microwave heating decrease the bioaccessibility of zinc while iron bioaccessibility is significantly enhanced in cereal grains including wheat.

Chapter 6

Chapatti Quality

Introduction

Chapatti is a flat, circular unleavened product made by rolling a ball of stiff dough of whole-wheat flour with or without the addition of salt and fat, and baked on both sides on a hot iron griddle popularly called *tawa*. Upon baking, it gets slightly puffed and acquires a golden brown colour with a few brown spots here and there on the surface. Chapatti is also commonly called *roti* and the two names of the product are synonymous. The chapatti though a baked product on its own right is categorized among flat breads both leavened and unleavened (Faridi, 1988; Qarooni et al, 1992). It has, however, no similarity with bread or cake except that both are made from the same cereal, wheat. Bread is essentially made from fine wheat flour (*maida*) of low extraction (70 – 72 %), whereas chapatti is essentially made from whole wheat flour of 95 – 100 % extraction. Bread comprises mostly crumbs with little crust, whereas chapatti has more crust and fewer crumbs; thinner the chapatti lesser the crumbs. Thus, the crust to crumb ratio in chapatti is much higher than that in bread. However, the quality of both bread and chapatti is determined using such parameters as physical, chemical, nutritional, microbiological and sensory, which are acceptable to the consumer, and meet the food grade standards. Nevertheless, the desirable characteristics of chapattis vary from culture to culture, region to region and necessarily from one household to another owing to differences in preparation procedure, and raw materials used. However, a soft and pliable texture, creamish colour, pleasing wheaty aroma and sweet taste along with good puffing are recognized as desirable quality attributes of good chapattis everywhere (Austin and Ram, 1971). But, consumers who hardly can afford a square meal to satisfy their hunger will perhaps be least bothered about these quality traits. It is the well- fed who look for certain quality traits in the chapatti. The chapatti quality is, therefore, the domain of a connoisseur.

Physicochemical Quality Characteristics of Chapatti

(See **chapters 1 & 5**).

Sensory Quality

The consumer ordinarily attaches more importance to the sensory quality. The sensory parameters that commonly define chapatti quality comprise visual appearance (such as shape, size, thickness, gloss and colour), puffing (partial, full), texture (soft and pliable, easy to tear, bite and chew), palatability or mouth feel, taste (sweet) and aroma (pleasing wheaty).

According to Yasin et al (1965) the criteria of chapatti quality should relate to stiffness, staleness, brittleness and pliability of chapatti, that is the chapatti should be soft not hard and brittle, fresh-like not stale, and pliable. However, Chaudhri and Muller (1970) used the following criteria of chapatti quality. The colour should be creamy white, recommending use of white wheat varieties and avoiding the use of red wheat varieties. The chapatti should be soft to feel and pliable. The taste of chapattis should be clean and untainted, and free of off flavours. When rolled the chapatti should be flexible and behave like a Swiss roll without any apparent signs of cracking. The chapattis should not be tough upon chewing. The drying rate of the chapattis should be slow after baking and upon storage since a hard chapatti is difficult to fold and chew, and is no longer acceptable to the consumer.

Austin and Ram (1971) defined the quality of chapatti from such determinants as water absorption in making dough, puffing, appearance, texture, palatability, aroma and keeping quality of chapatti. While evaluating the chapatti making quality of different wheat varieties/ cultivars, they observed that the whole wheat flour (*atta*) preferred for chapatti making purpose should have good water absorption and moisture retention capacity. Puffing of chapatti should be rapid and complete. The chapatti should have an attractive appearance, creamish in colour with soft, silky, smooth and pliable texture. The chapatti should be sweetish in taste with a pleasing aroma. The chapatti should retain these properties for a longer period.

Sinha (1973) listed the following quality aspects of a chapatti. The first thing that a chapatti-eater desires is that the colour of the chapatti should be neither red, nor brown, nor yellow, but of creamy-yellow to white colour, relatively more of the whiter shade. It should not be leathery or tough in texture and not too brittle either. Neither too much of grittiness nor stickiness is liked by the chapatti-eaters. It should not be either too dry or too soft and sticky to taste even if it is eaten some hours after its preparation. However, it should retain the normal moisture level. To certain extent, the chapatti should have a perceptible sweetish taste. Chapatti must puff well and fully while preparing so that the inner layer too is well baked and desirable texture is developed. Chapatti with slight caramel flavour and a few stray dark spots due to close contact with the glowing fire while preparing, is liked by many a consumer.

According to Mason and Hoseney (1980) the following quality characteristics are considered desirable. The chapatti should be uniform in diameter and thickness, well puffed throughout, and well baked inside. It should acquire tan to golden brown surface colour. The chapatti should be flexible and have a soft, silky appearance. The mouth feel of chapatti should be smooth and not gritty, and it should be easily chewed.

Ram and Nigam (1981) evaluated the chapattis using such quality criteria as puffing, pliability, hand tearing, taste and mouth feel during chewing.

According to Rao et al (1986b), the chapatti of good overall quality should possess an appealing colour with light brown spots spread evenly over the surface, a smooth, soft, and pliable hand feel, and the desired soft chewing and sweetish taste. It should be optimally baked and puffed so as to impart a pleasant wheaty aroma.

In summary, the baked chapatti should, therefore, conform to the following quality parameters for the satisfaction of the consumer:

- **Appearance:** The chapatti should have an appetizing appearance. It should be uniform in shape, size, and thickness, free from defects. The surface of chapatti should be smooth, untorn with very small golden brown spots spread evenly. The overcooked charred spots make it less appealing to the eye.
- **Puffing:** Puffing of chapatti improves appearance and consumer appeal. Consumers generally link full puffing of chapatti to its quality in terms of softness and pliability. The chapatti that puffs incompletely with scattered discontinuous puffs registers low consumer appeal (Austin and Ram, 1971). The height of puffed chapatti and pliability are reported to serve as simple indices of chapatti quality (Rao *et al*, 1986b). Puffing of chapatti is nothing but the separation of two distinct layers of the baking dough sheet inflated like a ball with pressurized steam in between generated due to flash evaporation of moisture in the dough. Puffing of chapatti may take place on the hot plate (*tawa*) itself or it may be effected by placing the partially dehydrated dough sheet on live charcoal or LPG flame or in suitably heated oven. The first partially baked side of the dough sheet should always face the hot surface/flame or live charcoal while puffing. When a chapatti puffs up rapidly into a completely inflated ball-like form, it is designated as "rapid and full puff." On the other hand, slow and incomplete puffing is described as "gradual and partial."At times chapatti does not puff at all, but has small discontinuous puffs scattered all over (Austin and Ram, 1971; Ram and Nigam, 1981). Such chapattis register low consumer appeal. Whole wheat flour that possesses good water absorption and moisture retention capacity yields well-puffed chapattis. More importantly, the chapatti dough should have a gluten matrix which is capable of retaining more steam resulting in higher puffed height. Rao et al (1986b) reported that poor puffed height of chapatti is attributed to poor quality gluten. Complete puffing is advantageous as it renders the inner layer, i.e. the crumb well baked due to the accumulated superheated steam and results in development of desirable texture and aroma in the chapatti. Dough mixing time, the thickness of dough disc/sheet, baking and puffing time and temperature are critical as slight variations greatly influence the chapatti quality, particularly with respect to puffing height, colour and pliability (Rao et al, 1986b). The height of puffed chapatti was found to increase with increase in the amount of water added to the dough as well as the temperature of puffing. Also, maximum puffing was observed in the case of thinner dough sheet/disc as compared to thicker one (Rao *et al,* 1986b). The temperature of the baking griddle (*Tawa*) and time for baking the first side of the dough sheet/disc are

also critical as over baking of the first side causes excessive dehydration leaving less moisture for steam formation, and under baking results in bursting because of decreased firmness of the sheeted dough disc surface. The change of colour of the top surface of the baking disc from creamish-white to wheatish with tiny blisters is the critical indicator to turn the first side.

- **Colour:** Colour is a key indicator of food quality. The colour of chapatti should be creamy-yellow to white with wheatish tinge. In Indian subcontinent lighter coloured or whitish *atta* and chapatti are preferred. The consumers generally do not like reddish, brown or yellow coloured chapattis. The wheats having amber colour are generally preferred for chapatti making as these yield creamish atta and chapatti (Misra, 1998).

  Since crust constitutes a major part of the chapatti its overall appearance in terms of colour and uniformity are the major factors governing chapatti quality.
- **Taste:** The chapatti should have perceptible sweetish taste. The whole wheat flour that contains sufficient amount of sugars, naturally present and those contributed by diastatic activity in dough, yields sweetish chapattis.
- **Flavour and Aroma:** Chapatti flavour is judged mainly in terms of sweetish taste and fresh typical wheatish aroma. A pleasing wheatish aroma is highly preferred. The non-enzymatic browning reactions occurring during baking and puffing too perceptibly contribute to the overall flavour of chapatti. Many chapatti consumers like the slight caramel flavour and a few stray very small dark spots that develop due to close contact with the glowing fire during the act of puffing (Sinha, 1973). The dark coloured spots on the surface of chapatti are due to the higher diastatic activity, which helps in the formation of more sugars during mixing of dough and subsequent operations, thus contributing to increased browning reaction.
- **Texture:** Ideally chapatti should be smooth, soft, silky and pliable so that it is easily torn and manipulated in hand to make a small scoop for holding curried preparations used as accompaniments. Chapatti possessing moderate tearing strength and more pliability requires lesser effort during chewing. It should retain these characteristics for a longer period as chapattis are more often eaten after 2 to 4 hours of baking (Rao, 1982). The chapattis that are dry, hard and brittle or tough and leathery are less pleasant to chew. Such chapattis are also difficult to break or tear and fold into scoop to hold curried preparations. According to Austin and Ram (1971) good quality chapatti should give minimum resistance to tearing.
- **Palatability:** Mouthfeel, palatability or eating quality is a sensory observation experienced during biting, chewing and eating of chapattis. It implies easy chewability with least mastication giving a feeling of easy dissolution in the mouth without any stickiness or grittiness.
- **Keeping-quality:** Keeping-quality of chapatti is equally important to the consumer since, more often, chapatti is consumed after 1 to 2 hours of baking.

It should retain all the above stated characteristics to the satisfaction of the user. Well-baked and well-packaged chapatti must retain freshness during the designated period of storage. Chapattis for the purpose of marketing must also satisfy the food grade standards and regulations to safe guard the interests of the consumer.

Determination of Chapatti Quality

The quality characteristics of chapatti enumerated above are mainly governed by the quality of wheat used and the processing conditions employed for converting it into flour. Since wheat quality and consequently the chapatti quality are governed by interaction of components during processing, a single test is insufficient to judge the quality of chapatti. Therefore, multiple tests are required to ascertain the chapatti quality attributes. These may be grouped under physicochemical tests, subjective, and objective tests. The physicochemical tests determine the composition of wheat, whole wheat flour and consequently of chapatti. The subjective tests determine the sensory attributes of chapatti. The objective tests, on the other hand, measure mainly the colour and texture of chapatti using instruments. Earlier studies largely relied on some physical, chemical, and sensory evaluation methods to judge the quality of chapattis (Austin and Ram, 1971).Several instrumental methods have since been developed, largely to estimate the colour and different texture related sensory attributes of chapatti. The objective evaluation using instruments requires less time, eliminates variations due to human judgment, and hence can replace the time consuming sensory evaluation procedures.

Physicochemical Tests

The physicochemical characteristics of wheat, whole wheat flour and that of chapatti may be determined using AACC (2000) and AOAC (2000) methods.

Sensory Tests: Sensory evaluation is a scientific method used to evoke, measure, analyze, and interpret those responses to food products as perceived through the senses of sight, smell, touch, taste, and hearing (Institute of Food Technologists, USA, 1981). Thus, the consumer liking for chapatti may be measured from the response a chapatti produces in consumer's senses of sight, touch, smell and taste after receiving and eating. For this purpose we need to take the opinion of a group of tasters since the sensory discrimination varies with individuals even though they are regular consumer of chapatti. There is, therefore, need to select an analytical taste panel. Those tasters who prove to be discriminating and consistent in their judgments are selected to take part in the taste trial. In addition to these attributes the tasters should preferably be conscientious and have real interest in the product quality assessment. Keenness often develops with practice but if the attitude of the taster is not one of interest in the activity then his/her judgments tend to be unreliable. And also that a panel selected by using one foodstuff will not necessarily show similar interest in a different food product. Therefore, only those persons, male or female who are regular consumer of the food product under test should be selected and preferably suitably trained to get reproducible results. To ascertain the opinion of different tasters various methods are used such as **discrimination test**, **descriptive analysis**, and **acceptance** (degree of liking, e.g. **hedonic scale**) or **preference test** (Amerine *et al*1965; Larmond 1977).Test method or design selected must ensure accuracy, and the results must be interpreted statistically.

The chapattis variously prepared in different laboratories have been evaluated for the above stated quality parameters to either determine the chapatti making quality of different wheat varieties or to study the effect of different additives to improve the sensory as well as nutritional value of the chapatti.

Austin and Ram (1971) while evaluating the chapatti making quality of different varieties of wheat, used scoring, grading and fractional pairs methods to evaluate the palatability (texture, taste and flavor) of chapattis employing tasters who were regular consumer of chapattis. In scoring method the palatability of chapattis prepared from wheat varieties under test was compared with that of chapatti prepared from a standard variety. A score of 10 was assigned to the palatability (sweetishness) of the standard variety chapatti. The chapattis of wheat varieties under test scoring lower or higher than 10 percent of standard score were adjudged inferior or superior respectively to the standard variety chapatti. In the grading method the selection of judges was done by providing chapattis made from one variety to 10 tasters simultaneously to test the palatability. The grades **(A+)** for sweetishness, **(A)** for moderately sweetish taste and **(A-)**for tastelessness were assigned. Those who gave **A+** and A were selected as judges for the palatability test. The method of fractional pairs in paired comparison for the evaluation of the palatability of chapattis was developed by Sadasivan *et* al (1971). The selection of judges was done first by conducting the *duo test* where a taster is asked to discriminate between two coded samples to eliminate biasness. First the tasters were given chapatti of the standard sample to acquaint them with the taste of the variety. Chapattis of different varieties along with standard variety chapatti were made into fractional pairs coded as AB, AC, and AD of which A is the standard. The pairs were given to tasters in random order and the samples from a pair were also at random. For every pair a judge was required to record the value 1 for the sample preferred and 2 for the other sample in the score card designed for the experiment. From these observations the ranking and matrix of preferences can be worked out. From these results we can know the number of times chapatti of standard variety is preferred to chapattis of other three varieties and how many times the standard variety chapatti is not preferred to chapattis of test varieties. The fractional pairs method was later modified using the Thurstone-Mosteller model so that splitting is symmetrical and each variety occurs equal number of times (Sadasivan *et al* (1976).

Chopra and Bhat (1975) evaluated the sensory quality of chapattis both fresh and those kept for four hours at room temperature using a panel of six judges. The acceptability of chapattis was determined in terms of appearance, texture, taste, chewing/ biting and keeping quality allotting scores of 7, 6, 5, 4, 3, 2 and 1 depicting strongly favourable, favourable, mildly favourable, undecided, unfavourable, mildly unfavourable and strongly unfavourable, respectively. The judges were asked to rank the different characteristics according to their importance in order to nullify the biasness of judges towards a particular characteristic and for working out weightages. The percent weightage was multiplied with the marks awarded by the judges for the sensory attribute in question. Among the sensory attributes tested, taste was thus reported to be an important criterion in evaluating the acceptability of chapattis.

Ram and Nigam (1981) used a panel of four trained tasters to assess the quality of chapattis in respect of pliability, hand tearing, mouth feel during chewing and taste. An arbitrary eight point scale, e.g. Excellent7- 8, Good 5-6, Fair 3-4, and Poor 1-2 was used to grade each quality criteria. The marks thus assigned were added and chapattis prepared from different wheat varieties graded as Excellent, Good, Fair and Poor.

A descriptive score card was designed to quantify different quality attributes of chapattis using a panel of six judges (Rao, 1982 and Rao *et al*, 1986b). The maximum possible score for each attribute was: height on puffing 10, pliability 10, appearance 20, hand feel 10, texture 25, and taste 25, so that the maximum total score would equal 100. Magnitude test was then used to ascertain the overall quality of chapattis. A chapatti receiving a score of above 70 out of total score of 100 was considered of good quality. The panel members were also asked to grade the overall quality of chapattis as excellent, good, fair, satisfactory or poor on the basis of sensory parameters evaluated.

Dhaliwalet al (1996) conducted the sensory appraisal of chapattis prepared from Canadian wheat varieties using seven trained panelists who assigned scores on the basis of colour 20, taste 25, aroma 20, chewiness 15, tearing ease15, and stickiness5. A score of 100 was assigned for overall appeal. Total scores by all panelists were averaged. Chapattis receiving more than 70 points were rated excellent, 60-69 points rated good, 50-59 points rated fair, and those receiving < 50 were rated unacceptable.

Rehman *et al* (2007) while evaluating chapatti making quality of British wheat cultivars in comparison with a commercial chapatti flour milled from a mixed grist of Australian wheats, employed 12 assessors (6 males and 6 females) who were regular user of chapattis. The panel members were briefed individually regarding sensory characteristics of chapatti and use of the score card. Chapattis were scored for colour 20, taste 15, aroma 15, breakability 20, chewability20 and stickiness10. Assessments were done in duplicate. Total scores of all panelists were averaged and chapattis were assigned 100 points for their overall appeal. Chapattis receiving above 70 points were classed as excellent, 60-69 points as good, 50-59 points as fair and those receiving less than 50 points were rated as poor.

In a number of studies on chapatti, a descriptive score card was designed to quantify different quality attributes (e.g. appearance, hand feel, colour, aroma, taste, texture, chewiness, tearing ease, breakability, stickiness, etc.) of chapatti using selected taste panel (Rao et al, 1986b; Dhaliwal et al, 1996; Hatcher et al,1997; Srivastava et al, 2002). The maximum possible score for each quality attribute was assigned in such a way so that the maximum total score would be equal to 100. Further, a number of studies used the Hedonic scale (9, 7, or 5 point) to evaluate the sensory quality of chapatti (Sadana et al, 1989; Deshpande et al, 2001; Rehman et al, 2006c; Ghodke and Ananthanarayan, 2007; Shaikh et al, 2007; Yadav et al, 2008a, 2008b, 2009b). Shaikh et al (2008) used 10 taste panelists to evaluate the texture of chapatti in terms of very soft, soft, slightly soft, hard and very hard assigning scores 9-10, 7-8, 5-6,, 3-4 and 0-2, respectively. An average score of 10 judgments was then calculated.

Objective Tests

- **Measurement of Chapatti Colour:** The colour of food surface is the first sensation that the consumer perceives and uses it as a tool to accept or reject the food. However, visual colour judgment can be affected by a number of factors, e.g. lighting conditions, angle of observation, individual differences in colour perception. The human eye is though capable to discriminate colour qualitatively, but is not a fine quantitative instrument. Consequently, precise colour measurement of foods for more consistent results has necessitated the need for modern instruments. The systems commonly used for colour measurement are the spectrophotometric and the Munsell systems. Each of these systems is a tri-stimulus system in which colour is specified by three attributes; dominant wavelength, purity, and lightness in the spectrphotometric systems; hue, chroma, and value in the Munsell system. Reflectance meters, and colorimeters / spectrocolorimeters have been used to measure the colour of chapattis (Larmond, 1977).

  Abrol (1972) studied the discoloration of chapattis using a reflectance meter. Readings were taken on the inside of the bottom layer of chapatti as the colour of the puffed top surface or outer bottom surface being uneven because of burnt or dark brown spots. Browning of chapattis upon heat sterilization was measured on a colorimeter after solvent extraction of chapatti (Kannur et al. 1973).

  Later studies to objectively measure the colour of baked chapattis used Hunter Lab Colorimeters (Hunter Lab, Reston, VA, USA).The values are recorded as $\mathbf{L}^*$ = lightness/brightness/or whitness (where 0 = black, 100 = white), $\mathbf{a}^*$($-a^*$=greenness, $+a^*$=redness) and $\mathbf{b}^*$($-\mathbf{b}^*$ =blueness, $+\mathbf{b}^*$ =yellowness) based on 1976 CIE L* a* b* color scale. These values are compared to a standard white plate. The surface colour of chapatti was variously studied including the effect of wheat variety and rate of flour extraction, effect of storage, effect of different additives to improve the sensory appeal of chapatti and prevention or retardation of discolouration (Hatcher et al, 1997; Shaikh et al, 2007; Hemalatha et al, 2007a; Ghodke et al, 2007; Yadav et al 2008a, 2008b and 2009a). Ebeler and Walker (1983) measured the inside colour of chapatti to eliminate the effect of residual flour used during rolling of dough sheet. The variety/cultivar of wheat and rate of flour extraction markedly affect the surface colour of chapatti. As the flour extraction rate increases flours tend to decrease in whiteness and increase in redness and yellowness Hunter values, consequently affecting the colour of chapatti accordingly. The consumer's liking for chapatti having creamish colour with a wheatish tinge, requires a chapatti to possess a balance of Hunter values in respect of whiteness, redness and yellowness. The chapatti made from resultant *atta* (mill *atta*), on the other hand, is more whitish (Rao et al, 1983).

- **Measurement of Chapatti Texture:** Texture is essentially the effect of physical and chemical properties on our senses of sight, hearing and touch (Matz, 1962). It is an important attribute in our perception of food if texture is not right; food is not liked even if the flavour/aroma is right. Evaluation of texture involves measuring the response of a food when it is subjected to forces such as cutting, shearing, compressing, stretching, chewing, etc. Giese (2003) had the opinion that "while sensory tests and descriptors are highly useful in understanding product rheology, there are problems and costs associated with training and maintaining of sensory panels. As an alternative, instruments for rheological food measurements are straight forward to use and provide consistent and significant data".

  Chapatti texture has several dimensions, e.g. flexibility / pliability, breakability / tear ability, and soft to touch, bite and chew. Puffing property of chapatti improves appearance and consumer appeal and projects it being soft and pliable (Rao, 1982).

- **Measurement of puffed height of chapatti:** Puffing quality is measured by height of puffed chapatti. The height of puffed chapatti in centimeter should be measured immediately after complete puffing using a centimeter scale. Navnidhi et al (2007) measured the height of the puffed chapatti using a specifically designed system having a ruler attached to it for convenience. The optimum height is considered to be 6.5 to 7.0 cm. Higher the puffed height more pliable the chapatti (Rao, 1993).

- **Measurement of chapatti pliability:** The pliability representing the softness of chapatti was measured by Rao et al (1986b) using a specifically designed apparatus. It is made of an iron channel with a vertically sliding clamp for fixing a strip of chapatti. A sliding graduated scale is fixed at the other end of the channel to measure the height of the bent chapatti strip. A chapatti strip measuring 2 x 7 cm is used to determine the pliability. Two centimeters of the total length of the chapatti strip is fixed between the two plates of the clamp and the remaining part of the strip is allowed to hang and bend under its own weight. The extent of bending is measured against the scale. The difference between the horizontal height at the center of the clamp, where the chapatti strip is fixed, and the height of the bent chapatti strip is taken as an index of pliability and expressed in centimeters. The pliability values obtained using this apparatus were found to be highly correlated to sensory evaluation data (Rao, 1982). A soft pliable chapatti is easy to fold into a scoop to hold curried preparations used as meal adjuncts.

- **Measurement of other chapatti texture indicators:** While eating chapatti, it is normally held in one hand and using the other hand a small piece of suitable size (1.5 to 2.5 square inch) is torn, folded into a scoop to pick curried preparation and then chewed in the mouth till it is pasty enough to swallow. This operation involves stretching, tearing and folding using hands and cutting and puncturing action of teeth in the mouth. The consumer prefers chapatti which is soft, easy to tear, flexible enough to be folded into a scoop as well as requires fewer efforts to chew. To simulate these actions closely tearing, stretching, extensibility, and cutting and puncture force measurements have been carried out using different devices attached to different texture measuring instruments. Warner- Bratzler Shear Press, Zwick Universal Testing Machine, Instron Universal Testing Machine, Stable Microsystems, UK, Universal Texture Measuring System- LLOYD Instruments, UK, General Food Corporation, USA, Stevens LFRA Texture Analyzer- Texture Technologies, etc. have been commonly used.

Khan et al (1972) used the precision penetrometer for determining the softness of chapatti. Rashid and Hawthorn (1974) conducted a number of tests with the Instron to measure chapatti texture indicators and found significant correlation between sensory panel scores and hardness or toughness of chapattis. Toughness is the work required to cause a rupture in a material. Instron Universal Testing Machine fitted with a sheer press apparatus having a plunger was used to measure shear force and toughness of chapattis (Ebeler and Walker, 1983; Gandhi and Bourne, 1988). Peak force and deformation to peak shear force data were used as an indication of toughness of chapatti.

The studies conducted by (Ram and Nigam, 1981; Rao, 1982; Rao et al, 1986b; Ranga Rao et al, 1986; Venkateshwara Rao et al, 1986; Rao et al, (1989) made use of Warner- Bratzler Shear Press, and the shear value was recorded in pounds after folding the chapatti into four layers. Higher the sheer force denotes harder the texture of chapatti while lower sheer force depicts that chapatti has softer texture, hence easy to tear and chew while eating. Rao (1982) reported a highly significant correlation between Warner- Bratzler shear values and sensory texture and overall quality of chapatti. However, according to Ram and Nigam (1981) the Warner- Bratzler Shear Press has limitations in studying wheat varietal differences because of lesser differences in the texture of different chapattis though the results are consistent with subjective evaluation. Later studies have used a variety of texture analyzers equipped with relevant attachments to study the above stated chapatti texture parameters.

Sidhu et al (1988) used Zwick Universal Testing Machine to measure stretching, tearing, cutting and puncture force using four different attachments. The two measurements tearing and stretching force simulate the force required by human hands to tear a piece of chapatti for eating, while cutting and puncture force represent the biting and chewing action of human teeth on chapattis. The extensibility of chapatti was measured in millimeters from the stretching force curve. Extensibility, which correlates with the flexibility and folding properties of chapatti, is reported to be one of the useful parameters in chapatti quality. Higher the extensibility more flexible the chapatti, and hence it is easy to fold and highly acceptable to the consumer. Hatcher et al (1997) performed the tearing / stretching test using a similar device attached to Instron Universal Testing Machine but instead used a trapezoidal piece of chapatti. For measuring total penetration a rounded plunger was used to perform puncturing force test. Shaikh et al (2007) instead used a needle probe attached to Stevens LFRA Texture Analyzer to penetrate chapatti. Texture was expressed as the load in grams required penetrating the chapatti.

Venkateswara Rao et al (1986) adapted the tearing resistance meter (Elmendor Paper Tearing Tester, UK) to measure the tearing strength of chapatti during storage.

Yadav et al (2009b) and Khan MA et al (2011), for measuring the hardness, cohesiveness, springiness, gumminess and chewiness attributes of chapatti, performed the bite test on chapatti using the volodke bite set (part no. 01/2663) attached to the Texture Analyzer Plus (Model No. 01/TALS/LXE/UK, LLOYD Instruments, UK). This bite set has been designed to simulate incisor teeth shearing through a food material. The force-time deformation curve derived from this test was used to calculate the following texture attributes for chapatti: hardness value as peak force of the first compression; cohesiveness as the area of work done during the second compression divided by the area of work done during the first compression; springiness as the distance (mm) the chapatti recovers after the first compression; gumminess as product of hardness x cohesiveness; chewiness as the product of gumminess x springiness. The chewiness attribute thus calculated equates the energy required to chew a solid food product in the mouth. Chapattis having lower springiness and higher chewiness values tend to be physically more rubbery and harder to chew.

Gujral and Pathak (2002), and Ghodke and Ananthanarayan (2007) tested a rectangular strip of chapatti for extensibility on the Instron Universal Testing Machine. The chapatti strip was pulled apart between two clamps until it ruptured. The maximum force of the force- displacement curve was the peak force; the distance to the peak force was the extensibility. The energy to rupture is the area under the force displacement curve and is dependent on the extensibility and peak force. The low values for peak force and the calculated modulus of deformation as well as by the longer distance extension reached before the rupture showed that the chapatti was soft and extensible, a quality well appreciated by the consumer.

Sharma et al (2004) standardized a method for measuring force required to tear chapatti using Instron Universal Testing Machine. In this method the chapatti was stretched until torn using Instron in the same manner in which chapatti is pulled between fingers by the consumers while eating. An elongation curve was obtained and the area under the curve was measured by planimeter and was expressed as Elongation Curve Area (ECA). ECA of chapatti showed inverse relationship with mouthfeel and overall acceptability. According to the authors, ECA of chapatti is a direct method to evaluate chapatti making quality of whole wheat flour.

Factors Influencing Chapatti Quality

Influence of Wheat Type and Variety

- ***Wheat type*:** Among the different*Triticum* wheat species, *T. aestivum* is most suitable for chapatti making because it yields non-sticky dough with good machinability, and a fully puffed chapatti with soft and pliable texture. The species *T. durum* and *T. dicoccum* are also sometimes used for chapatti making but result in dry and hard texture products which are perceived to be leathery when chewed and also sweeter owing to the higher starch damage during milling (Rao, 1993). Supekar et al (2005) also found *aestivum* cultivars superior in hectoliter weight, sedimentation value and chapatti making quality compared to durum cultivars. The durum cultivars, however, scored over *aestivum* cultivars in grain appearance score, thousand grain weight, contents of starch, crude fiber, crude protein, dry gluten, iron, zinc, β-carotene and water absorption capacity. Chapattis prepared from *aestivum* flours were superior to those made from durum and *dicoccum* species.
- ***Aestivum Wheat Varieties*:** The quality characteristics of chapattis are mainly governed by the quality of wheat used and the processing conditions employed for converting it into flour (Austin and Ram, 1971). The major grain characteristics that determine wheat quality are grain hardness, protein content, gluten strength, lysine, amylase and β-carotene contents, and components related to water absorption (Ram and Mishra, 2007). Those wheat varieties that are medium hard to hard with bold, sound, uniform and lustrous amber golden colour grains, are preferred for making good quality chapattis (Misra, 1998). Medium strong wheats with higher protein content and hard grain texture with high starch damage and high water absorption capacity are preferred for chapatti making (Srivastava et al, 2002).Soft wheats generally make poor quality chapattis.The wheat selected should also have low or no polyphenolase activity. It should be free from any infection and insect infestation as such wheats yield chapattis of inferior quality because of the development of extraneous colour and flavour during dough resting and subsequent baking (DWR Progress Report, 2008). Further, in selecting wheat variety for making palatable chapattis, initial sugar content and damaged starch should serve as useful index (Uprety and Abrol, 1972).Some of the promising varieties of *aesativum* wheat that make chapattis of good quality are listed in Chapter one.

  Several wheat varieties/cultivars from USA, Canada, Australia, and UK tested for chapatti making quality have not proven totally fit for chapatti of acceptable quality largely because these were developed for making bread and similar leavened baked products. However, Rehman et al (2006c, 2007) have shown that blending two British wheat varieties in appropriate proportion resulted in chapattis of soft texturewhich did not change significantly during storage. Studies carried out on interchangeability of flour constituents between good quality chapatti making wheat and poor quality chapatti making wheat have shown that interchanging gluten (Srivastava et al 2002), and arabinoxylans (Hemalatha et al, 2013) improved the chapatti quality of poor chapatti making wheat. Therefore, blending poor chapatti making wheat with good chapatti making wheat in requisite ratio is feasible and should be practiced to produce chapattis of satisfactory quality. Similar observations were earlier made by Ram and Nigam (1979) who worked out a ratio between residue proteins and gliadin + glutenin, which may be used as index of flour strength for the purpose of formulating flours of desired strength for producing baked products like chapatti of desired quality.

- ***Influence of Milling Method:*** Stone, disc, hammer, pin and roller mills have been tested to produce whole wheat flour. Stone milled (particularly water driven) whole wheat flour makes superior quality chapatti with soft texture and better wheaty flavour compared to flours obtained from other mills (Leelavathi et al, 1986). The quality of chapatti may be further improved by manipulating the stone *chakki* settings for improving the quality characteristics of whole wheat flour. (Inamdar and Prabhasankar, 2017).

Influence of Physicochemical Characteristics of Whole wheat Flour

Flour extraction rate, particle size, temperature generated during milling, protein quantity and quality, lipid content, sugar content and diastatic activity, damaged starch content produced during milling, polyphenolase activity, and rheological properties of flour all affect the final quality of chapatti.

(a) ***Influence ofFlour Extraction Rate*:** Traditionally, 100 percent extraction whole wheat flour is used to make chapattis with some households, however, sieving off 4 to 6 percent coarse bran before making dough, thus virtually using 94 to 96 percent extraction flour (Rao et al, 1986b).Sabir et al (1982) have shown that simultaneous removal of fines (0-40 per cent) and bran (0-10 per cent) results in highly refined flour containing lower level of mineral elements, which not only affects the nutritional value but also overall quality of chapatti. However, the combination of simultaneous removal of 5 per cent bran and 10 per cent fines yielded *atta* which possessed better nutritional, dough kneading, chapatti baking and puffing qualities. The practice of removing higher amounts of bran by sieving, or using a flour of a lower extraction rate makes the chapatti tough and leathery (Rao, 1993). In urban areas, mechanical abrasion of wheat prior to milling is commonly practiced, and variable percentage of coarse bran and endosperm are sieved off at the mill itself resulting in very fine whitish whole wheat flour of 80 to 86 per cent extraction, which the urban consumers prefer as it yields more whitish chapattis (Kaur *et al*, 1990).However, the abrasive removal of bran from wheat prior to milling results in not only considerable losses of essential nutrients (protein, minerals, fat, crude fibre, essential amino acids, etc.) but also affects adversely the textural characteristics and flavor of chapattis (Kaur *et al*, 1990 and Bajaj *et al*, 1990). Hatcher *et al* (1997) evaluated a wide range of extraction rate roller mill flours of Canadian wheats for their chapatti making quality and concluded that low-extraction flours were less suitable for chapatti making. The 85 per cent extraction flours, however, yielded the preferred chapatti having creamy yellow colour, wheaty aroma and taste, and soft, flexible texture with easy tearing.

(b) ***Influence of Flour Particle* Size:** The flour granulation or particle size also greatly influences the dough characteristics and the quality of chapatti (Sharma and Bains, 1976; Rao, 1982). The flours with smaller average particle size exhibit higher water absorption values because of greater specific surface, and higher damaged starch content. According to Aziz and Bhatti (1960) for chapatti to be of good quality, flour should be fine, about 90 per cent passing through 50 mesh sieve. The whole wheat flour obtained in water driven stone *chakki* contained 4.7 per cent overs on 355μ sieve and 72.6 percent overs on 150μ sieve and was found superior for making chapattis of good quality (Sharma and Bains, 1976). Rao (1982) used disc-milled whole wheat flour and reported that very coarse and medium coarse flours containing 77 and 67 per cent over tailings in 129μ sieve, respectively yielded chapattis with tough texture and poor overall quality, while chapattis made from normal or fine flours containing 31 to 54 per cent over tailings in 129μ sieve were of better quality with high acceptability. Islam and Johansen (1987) who used hammer mill, reported that whole wheat flour of particle size of 1000μ and below was satisfactory for making chapattis with desirable sensory characteristics. Sidhu *et al* (1988) reported that roller mill whole wheat flour having more of flour particles below 250μ produced chapattis of highly desirable texture. These findings have been corroborated by Rao (1993) that the fine - ground flour is desirable for chapatti making and the coarse flour with 70-80 per cent and more over tailings on a 129μ sieve makes chapatti having a poor appearance and tough texture.

(c) ***Influence of Flour Temperature***: Whole wheat flour milled particularly in disc or stone mills generally attains high temperatures ranging up to 95°C due to severity of grinding. Inamdar et al (2015) recently reported variations in flour temperature from *chakki* (75= 84°C), disc (69- 78°C), and hammer mill (67- 74°C) compared to roller milled flour (35- 38°C). The heat thus produced has been reported to improve the quality of chapatti (**Table 1**) particularly potentiating the wheaty aroma and taste besides having significant effect on lipids, protein fraction, and diastase and proteolytic activities, and lowering the total microbial load (Rao, 1982; Ranga Rao *et al*, 1986). However, Prabhasankar and Rao (2001) have shown that heat produced in stone and plate mills denatures the protein, and reduces the amino acids and the unsaturated fatty acids content of *atta*.

Table 1: Effect of heat developed during grinding wheat in disc mill on the quality of chapatti

Whole wheat Flour	Puffed chapatti height (cm)	Pliability (cm)	WB** shear value (lb)	Appearance	Handfeel	Texture	Flavour	Overall quality
Control*	6.0	1.8	8.2	Dull grey spots, non-uniform	Slightly rough, less pliable	Slightly rough	Bland	Satisfactory
Cooled immedi-ately	7.1	2.2	5.8	Light brown spots, uniform	Smooth, pliable	Soft	Wholeso me, wheaty	Good
Cooled slowly	7.2	2.1	6.0	”	”	”	”	”

Source: Ranga Rao et al (1986). *Ground in pestle and mortar. ** Warner Bratzler

(d) ***Influence of Flour Damaged Starch Content*:** More of damaged starch is produced during milling of hard to medium hard wheats as compared to soft wheats because of severity of grinding required to yield flour of suitable particle size. The extent of damaged starch depends upon the type of grinding device used, stone mills yielding the highest amount (18.8 per cent) while the lowest (8.2 per cent) was in roller mill flour (Leelavathi et al, 1986). Recently, Ghodke *et al* (2009) have shown that the milling conditions like narrower the mill aperture, slower the feed rate and higher the moisture content of wheat grain, result in flour of higher damaged starch content (26.9 percent). The damaged starch produced during milling has enhanced capacity to absorb more water and thus markedly influences the quality of chapatti in terms of texture, taste and flavour. The diastatic enzymes work at a faster rate on damaged starch producing maltose readily. Higher the maltose value, better the chapatti quality in terms of texture, flavour, and pleasant sweetish taste (Rao *et al*, 1989). However, the damaged starch in flour exceeding 16.5 per cent and excessive diastatic activity result in sticky dough which becomes difficult to roll or sheet, and the chapatti upon baking develops darker unappealing spots on the surface. Also, higher the amount of damaged starch in whole wheat flour, higher the amount of acrylamide formation during the baking of chapatti (Mulla et al, 2010). Acrylamide is suspected of having carcinogenic properties, an adverse effect of higher content of damaged starch in whole wheat flour.

(e) ***Influence of Flour Protein Quantity and Quality*:** Earlier studies indicated that wheat containing 10 to 13 per cent (Singh and Bailey, 1940), 9.5 to 10.5 per cent (Austin and Ram, 1971), and 10 to 12 per cent (Austin, 1978a) protein was suitable for making chapattis with desirable quality. Rao (1993), however, reported a protein content of 9.0 to 10.5 per cent being optimum for making chapatti of satisfactory quality. Medium- hard wheat containing more than 11.0 percent protein result in a chapatti having tough and leathery texture, whereas a soft wheat having less than 9.0 per cent protein yield a chapatti with a stiff and brittle texture. Such chapattis upon storage become hard and brittle within an hour. Subsequent studies, however, indicated that even wheat with a higher protein content (> 12 per cent) was also suitable for making good quality chapatti (Srivastava et al, 2003). Hemalatha et al (2007a), on the other hand indicated that the total protein content of wheat had no influence on the chapatti quality as such, rather characteristics of protein and other constituents of wheat may also contribute to the quality of chapatti. Manu and Prasada Rao (2008) reported that high proportion of sodium dodecyl sulphate (SDS) extractable large polymeric proteins in whole wheat flour protein increase the toughness of chapatti texture, while flours having high thiol content decrease the toughness of chapatti. This indicates that it is not only the quantity but also the quality of the protein that reflects the chapatti making quality of a wheat variety. It has been shown earlier that the qualitative differences in wheat proteins outweigh quantitative differences in determining baking quality (Dexter and Matsuo, 1978).

The quantity and quality of gluten, which is formed as cohesive mass upon mixing flour with water, play greater role in chapatti quality. Arya et al (2016) have shown that gluten fraction of whole wheat flour (WWF) plays an important role in deciding the final quality of dough and chapatti. These authors demonstrated that addition or deletion of gluten in the WWF by more than 1 per cent affected the chapatti quality. Thus it is possible to manipulate the sensory as well as textural properties of chapatti by varying flour composition through fractionation or addition or by complete reconstitution. The present study clearly indicated that gluten was the major constituent which imparted specific chapatti quality characteristics to a wheat variety. It has been reported that the viscoelastic property of dough, which influences the baking quality of wheat flour, depends on the quality and quantity of protein. Very high content of protein and gluten of high strength makes the chapatti tough and leathery. On the other hand, wheat with very low protein (< 9 per cent) and weak gluten makes crumbly type of chapatti having poor texture and keeping quality (Austin and Ram, 1971).The strength of gluten is measured by Zeleny Sedimentation and Pelshenke tests. The wheat varieties having sedimentation values and Pelshenke values between 30 to 37, 30 to 40, 25 to 35, 30 to 60 ml; and 120 to 150 min, 100 to 150 min, respectively are generally reported to yield chapattis of acceptable quality (Singh and Bailey, 1940; Austin, 1978b;

Rao, 1993; DWR Progress Report, 2008). The protein content, sedimentation value, and Pelshenke value influence the textural characteristics, while the sugar content and diastatic activity determine the taste and flavour of chapatti (Singh and Bailey, 1940). Srivastava et al (1998) reported the influence of gluten content and its composition in whole wheat flours from two different wheat varieties on chapatti quality. Interchanging the gluten isolated from good quality chapatti-making flour to poor quality chapatti-making flour substantially improved the quality of chapatti made from the latter, thus suggesting the role of gluten content as well as its composition. Srivastava et al (2002) have further shown microscopically that in hard wheat flour dough gluten strands being large cover the starch granules, while in soft wheat flour dough gluten is short and starch granules remain exposed. Consequently, higher moisture retention and gelatinization of starch in hard wheat flour result in pliable and soft textured chapatti.

The three constituents of gluten, namely gliadin, glutenin and residue proteins have distinct properties (Ram and Nigam, 1979). Glutenins regulate strength and elasticity whereas gliadins determine the viscosity and extensibility of the dough. The residue proteins contribute significantly to dough extensibility. The relative quantities of these three gluten components play an important role in determining the chapatti characteristics. It has been shown that wheat varieties having roughly equal quantities of these proteins will make better chapattis than varieties in which excessive amounts of one of the other gluten components are present. The residue proteins appear to play a greater role in chapatti quality (Srivasrava et al, 1998). Ram and Nigam (1981) have shown that both low and high proportions of residue proteins are detrimental to chapatti quality. Low proportion of residue proteins is associated with stiffness and partial puffing and high proportion of residue proteins with leathery texture and full but slow puffing of chapatti. The studies conducted by Rehman et al (1997) revealed that glutenin contributes strength and gliadin imparts weakness to wholemeal flours. Prabhasankar (2002) reported that wheat varieties with low gliadin content were more suitable for making chapattis. Sharma et al (2004), on the other hand, have shown the possibility of the glutenin fraction of gluten being important to the chapatti making quality of whole-wheat flour. These contradictory observations do not clarify which protein quality factors are important in judging the chapatti quality parameters. Since bread-making quality is highly influenced by the HMW subunits of glutenin (Payne *et al*, 1987), they may also play a role in determining chapatti making quality of Indian wheats. Srivastava *et al* (2003), while evaluating 15 Indian wheat cultivars differing in protein content and SDS-SV, observed that excellent chapatti quality was associated with the presence of HMW glutenin subunits 5+10. The subunits 2+12, on the other hand, were found to be associated with poor chapatti quality. Earlier, Anjum et al (2000), however, reported that subunits 5+10 or 2+12 did not influence the chapatti quality. Sreeramulu et al

(2004) have reported that wheat varieties possessing high molecular weight glutenin subunits (HMW.GS, 1B20) produced chapattis with higher consumer acceptance. Das *et al* (2006), however, did not find significant differences in chapatti making quality of wheat selections differing in specific HMW subunits of glutenin indicating that individual HMW subunits / pairs are less likely to be a major factor in influencing chapatti making quality. Ram and Mishra (2007) have demonstrated that HMW.GS is alone insufficient for differences in chapatti quality; LMW.GS must also be taken into consideration along with HMW.GS. Hemalatha et al (2007a) have reported that low molecular- weight protein (gliadin) content and high molecular-weight glutenin subunits (HMW.GS) composition influenced the chapatti quality. The low molecular-weight protein content significantly correlated to the overall quality of chapatti, while HMW.GS composition significantly correlated to the texture of chapatti. However, the peroxidase activity which had negative influence on the colour of chapatti resulted in reducing the overall quality.

Though it is believed that the content and quality of the protein in a wheat variety is of major importance in determining the baking quality of the flour milled from the wheat, the importance of the carbohydrate profile, the starch configuration, and the enzymes present in it cannot be overlooked (Jackel, 1987).

(f) ***Influence of* Flour *Sugar Content and Diastatic Activity*:** Sweetishness and flavour of chapatti are closely dependent on the amount of sugar originally present in the flour combined with the sugar produced through the hydrolysis of starch by the diastatic enzymes when the dough is made and kept for sometimes before chapattis are made (Singh and Bailey, 1940; Abrol and Mehdi, 1971). The whole wheat flour commonly used for chapatti making is reported to contain total sugars in the range of 2.5 to 4.0 percent (Rao, 1993). Extensive chapatti palatability tests have shown that this range of sugars is in fact needed to make chapatti sweetish and palatable to the ordinary chapatti eaters. Additional amount of sugar is produced in dough by the action of amylolytic enzymes on the damaged starch produced during milling. Diastatic activity between 150 to 375 mg per 10 g in whole wheat flour is considered suitable for chapatti making (Singh and Bailey, 1940; Rao, 1993).It is, therefore, necessary that *atta* used for making chapattis should have desirable combination of total sugars and diastatic activity. Accordingly, the wheat varieties having suitable combination of these characters should be selected for chapatti- making. The whole wheat flour that contains less than 2.5 per cent total sugars, yields chapattis that score poorly for palatability and are more often judged insipid (Austin and Ram, 1971). The excessive diastatic activity, on the other hand, results in sticky dough and chapatti upon baking develops darker unappealing spots on the surface. During short period of baking at temperatures over 200^0C, caramelization and Maillard reactions contribute to the flavour of chapatti and bring about dullness of chapatti surface and formation of dark brown or burnt spots (Abrol et al, 1972b; Kannur et al, 1974). However, these reactions are not responsible for browning of chapattis. The phenolic constituents of the wheat grains coupled with their high tyrosinase activity result in browning observed in chapattis prepared from some of the wheat varieties (Abrol et al, 1971).

(g) ***Influence of Wheat Flour Lipids*:** Wheat grain contains 2 to 4 per cent lipids of which bound lipids are embedded in starch granules and are effectively inert. The free lipids are the one that participate in physical, chemical and biochemical processes in flour and dough. Lipids play a multifarious role in baking. They modify gluten structure at dough mixing stage, prevent interaction between starch granules during gelatinization, retard water transport from protein to starch, retard starch gelatinization, act as antistaling agents, etc. (Pomeranz, 1988). The polar wheat lipids play an important role in determining the chapatti making characteristics of whole wheat flour (Arya et al, 1978). The removal of petroleum ether soluble lipids from atta does not influence the texture hardness of chapattis but removal of total lipids adversely affects the texture of chapattis. The occurrence of high amounts of polar lipids in whole wheat flour results in softer chapattis (Rao, 1993).

(h) ***Infuence of Polyphenolase Activity of Wheat*:** The wheat variety that gives low response to phenol test possesses low polyphenolase activity and is considered of good chapatti making quality (DWR Progress Report, 2008). The indigenous wheat varieties are reported to be low in these enzymes, whereas the imported Mexican hybrid wheat varieties now grown in India have a high content of these enzymes. The adverse effect of enzymatic browning is greater when the chapatti dough is rested for a longer period, and also when the baked chapatti is stored for a few hours (Rao, 1993).

(i) ***Influence of Rheological Properties of Flour*:** See Chapter two for details.

Influence of other Ingredients

(a) ***Influence of Added Salt:*** The use of edible common salt in chapatti dough is optional. Nevertheless, when used at 1.0 or 1.5 per cent level the eating quality of chapatti is reported to be optimum with slight softening of texture. The dough containing 1.5 per cent salt was relatively softer to hand feel, cohesive, non-sticky and had better rolling/ sheeting characteristics. On the other hand, dough containing no or as little as 0.5 per cent salt was short and weak, with poor rolling characteristics, and the chapatti was flat in taste (Shurpalekar and Prabhavathi, 1976). According to Rao et al (1986a), the addition of salt up to 2 per cent level increased the extrusion time due to stiffening action of salt on the dough, but improved the cohesiveness and rolling properties without any stickiness. Majeed et al (1987) reported that addition of 1.0 or 2.0 per cent common salt increased the water absorption from 65.5 to 69 per cent and dough development time from 9.5 to 10.0 and 12.5 per cent, respectively. Raghavan et al (1994) reported an improvement in the sheeting characteristics with the addition of 1.5 per cent salt to the whole wheat flour dough. Addition of salt, however, reduced the farinographic consistency but caused an increase in dough development time, resistance to extension and stability as well as elasticity of the dough. Gill et al (2005) used 0.5 to 2.0 per cent sodium chloride in chickpea supplemented whole wheat flour chapatti and found improvement in its taste and mouthfeel.

(b) ***Influence of Added Oil or Fat*:** The use of refined edible oil or hydrogenated fat or *ghee* in chapatti dough is optional. However, incorporation of edible oil or fat mellows the dough improving thereby the consistency, handling and rolling properties. The addition of 6 per cent fat is reported to decrease extrusion time from 63 to 56 s (Rao et al, 1986a). Addition of oil in chapatti dough improves the texture of chapatti, and also keeps the chapatti soft and pliable during storage. Incorporation of very small amount of oil in the dough also improves the puffing characteristics of chapatti; however, higher levels adversely affect the puffing quality of chapatti (Rao, 1993).

(c) ***Influence of Water for Preparing Chapatti Dough*:** The water used for preparing chapatti dough should be potable having medium hardness, and a pH close to neutrality (Rao, 1993). The quality of chapatti is reported to be influenced by the dough consistency, which in turn depends mainly on the quantity of water added (Austin and Ram, 1971).The preparation of chapatti dough of desired consistency has relevance in the commercial production of chapatti and selection of new wheat varieties for their chapatti making quality. The amount of water needed to prepare dough of optimum consistency for the preparation of chapatti of satisfactory quality varies from 65 to 75 per cent (Rao, 1993). The amount of water absorbed depends on several factors such as the type and variety of wheat, milling method, rate of extraction, particle size distribution, severity of grinding, etc. The higher the absorption and retention of water in the dough, the softer will be the texture and sweeter the taste of chapatti. Rao (1982) observed that the chapatti dough made using 10 per cent more water than the normal amount, mixing for 1 min and without resting, were the best conditions for the preparation of chapatti. This will not only keep the chapatti soft for longer period but also minimize the power consumption and increase the yield of chapatti particularly in the commercial preparation and marketing of the product.

(d) ***Influence of Additives*:**In India different wheat varieties / cultivars are produced and quality of chapatti varies depending on the variety. The consumer faces the problem of availability of right variety of wheat flour suitable for chapatti preparation. The use of unsuitable wheat flour for chapatti preparation affects the quality of the product resulting in a tough, non-pliable, fragile and undesirable chapatti. The reduction of chapatti hardness is vital because the quality and price of the final product depends on this parameter especially when marketed in unit packs. Food additives are commonly used to improve the texture and keeping quality of baked food products. Mechanized production and increased consumer demand for high-quality chapatti with long shelf-life has, therefore, necessitated the use of functional food additives. The food additives that have commonly been used include: non-fat-dry-milk (NFDM). yogurt, butter milk, yeast, lactic acid, ascorbic acid, emulsifiers or surfactants, e.g. glyceryl monostearate (GMS), sodium steroyl -lactilate (SSL), sucrose esters; enzymes (α-amylase, xylanase); hydrocolloids such as guar gum, carboxymethylcellulose (CMC), hydroxypropylmethylcellulose (HPMC), k-carrageenan; and several other additives.

Hussain and Satti(1969) reported that the use of GMS and added vital gluten in dough kept the *roti* soft and resilient for up to 72 h at 35°C. The addition of GMS and NFDM slows down the loss of moisture making chapattis / *rotis* retain freshness for longer period. Ascorbic acid within the concentration of 0.1 to 0.2 per cent (on whole wheat flour basis) was found to prevent discoloration of chapattis, and improved the texture and puffing of chapattis those prepared from *resultant atta* (Abrol et al, 1972a). Addition of *vanaspati* (hydrogenated fat), and mono-glycerides has been found beneficial in the retention of soft texture in preserved chapattis (Arya et al, 1977). Sorbic acid generally used as preservative when added to dough, was found to decrease dough development time and tolerance of dough but did not cause any change in sulphydryl groups of proteins, and did not affect the quality of chapatti (Vidyasagar and Arya, 1981).Addition of yeast along with sugar, fat, GMS, SSL, lactic acid, and α-amylase to dough improved the texture and sweetness, and reduced the phytates content without affecting the consumer acceptance of chapattis (Swarnjeet et al, 1982).Peak gelatinization viscosity increased while amylose solubility decreased as a result of addition of fatty acids, mono- and di-glycerides and all the surfactants into *atta* used for chapatti making (Arya and Murthy, 1983). However, guar gum, locust bean gum, carrageenan and CMC were not found effective. The use of shortening (partially hydrogenated peanut oil) at 3 per cent level in combination with GMS (0.5 per cent) produced chapattis which were soft and flexible when fresh and showed the least amount of firming during three days storage at 20ºC (Sidhu et al, 1989). Chapattis prepared from pH adjusted (lactic acid) yeasted dough and rested for 4 to 8 h were softer than the control fresh chapattis. The addition of lactic acid softened chapattis more than addition of yeast (Syed et al, 1991).Chapattis made from whole wheat flour with 0.5 per cent GMS were softer and more desirable in appearance, and possessed wheaty aroma and resisted tearing during storage (Pirayeshfar et al, 2001). Manohar and Rao (2005) have developed and patented a novel improver premix comprising ascorbic acid, sodium srearoyl-lactilate and guar gum for improving the quality of chapatti with longer shelf life. While investigating the effect of freeze-thaw cycles on rheological and sensory quality of ready-to-bake-frozen chapattis, Yadav et al (2008c) reported that addition of glycerol (1%) improved the texture and mouthfeel of chapattis during freeze-thaw cycles. Shaikh et al (2008) while investigating the effect of addition of SSL, GMS, propylene glycol, sorbitol, α-amylase, xylanase, maltodextrin and guar gum to chapatti dough on the storage quality of chapattis at 29 ± 1⁰C and 4 ± 1⁰C reported that all the additives inhibited staling of chapatti to different extent. However, maltodextrin (0.3 % w/w) and combination of α-amylase (100 ppm) and SSL (0.375%) were found best to extend the storage life of chapattis. Glutathion (100 ppm) and L-cysteine (90 ppm) added to whole wheat flour prevented discoloration of flattened circular dough during refrigerated (5- 6⁰C) storage for three days (Yadav, et al, 2009a). The effectiveness of these chemicals increased in combination with amino guanidine bicarbonate (0.5 per cent), a compound known to retard the Maillard browning reaction. The baked chapattis were overall acceptable. Additives like wet gluten, liquid shortening, GMS, CMC, sodium caseinate and skim milk powder were found to significantly increase the extensibility of chapatti and kept the chapatti more extensible even after 24 h of storage at 25⁰C (Gujral and Pathak, 2002). The enzyme diastase, however did not significantly affect the extensibility of fresh chapatti but the chapatti remained more extensible than the control chapatti after 24 h storage.Gujral and Gaur (2002) showed that barley flour, wet gluten and liquid shortening in various combinations markedly improved the quality of fresh as well as stored (24 h) chapatti The hydrocolloids (guar gum, CMC,HPMC, k-carrageenan) added to chapatti dough improved the chapatti quality; however, the highest improvement in overall quality of chapatti was brought about by guar gum and HPMC when added at 0.75 and 0.5 per cent level on flour weight basis, respectively (Ghodke and Ananthnarayan, 2007). The chapattis made with these hydrocolloids possessed attractive surface characteristics, excellent softness and pliability. Khan AN et al (2011) fortified wheat flour with CMC, ascorbic acid, lecithin and sodium propionate to study the effect on dough rheology along with baking and sensory attributes of chapattis during storage for 24 h. The additives were found to improve the pasting properties, water absorption capacity, and dough development time and dough stability of the wheat flour. The addition of additives improved almost all the sensory attributes (colour, taste, aroma, texture, chewability, foldability and overall acceptability) of chapattis as compared to the control. The chapattis prepared from dough treated with enzymes (α-amylases- bacterial and fungal, and xylanase) had soft texture, better pliability and higher overall quality scores compared to the control (Hemalatha et al, 2010). Also, chapattis prepared from whole wheat flour dough treated with amylases and xylanase contained increased amount of soluble dietary fibre and soluble polyphenols with significantly improved antioxidant properties as compared to the control (Hemalatha et al, 2012). Hemalatha et al (2013) further reported that arabinoxylans- the non-starch polysaccharides isolated from good chapatti making wheat varieties were found to improve the chapatti quality of poor chapatti making wheat flour. The chapattis had soft texture and high overall sensory quality scores. The chapattis prepared by Suribabu et al (2014) from wheat flour dough mixed with 100U / mg protein of α-amylase (from *Brevibacillus borstelensis* R1) possessed improved taste, flavour, crust colour, softness and toasting qualities.

Influence of Chapatti Making Procedure

Chapatti making involves preparation of dough, sheeting of dough, baking of dough disc, and puffing of the baked disc. All these steps are critical requiring optimization of conditions. The preparation of dough of desired quality requires the addition of optimum quantity of water, and mixing for optimum time in a mixer or by manual kneading followed by resting suitably. A good quality chapatti dough should show optimum values for the following characteristics: consistency, stability, elasticity, cohesiveness, extensibility, and resistance to extension, adhesiveness as well as the strength of dough (Rao, 1982). The optimum values of these characteristics determine the ease of rolling / sheeting, formation of continuous dough sheet without breaking, the retention of the desired shape of the rolled dough disc before transfer to the baking griddle *(tawa)*.It is desired that the rolled dough disc should be of uniform thickness all over so that the baking is uniform and the chapatti puffs evenly without any rupture. The baking of the dough disc should be carried out at an optimum temperature for optimum time on both sides to curtail under baking or over baking. The first side of the baking disc should be turned upside down as soon as the top surface assumes wheaty tinge and develops tiny blisters, otherwise the chapatti will not puff fully. The under baked first side will rupture when puffed because of the weak raw surface, while the over baked first side results in poor puffing because of decreased moisture content needed for sufficient steam formation to inflate the chapatti. Puffing of chapatti if done in a preheated oven, time and temperature should be optimized, otherwise at low temperature it will not puff well and may become stiff; high temperature may cause charring. Puffing on the open flame, on the other hand, requires experience, skill and extra care to prevent burning and hardening of the crust. The optimum parameters for making dough, baking and puffing of chapatti of desired quality are described in Chapter 4.

Chapter 7

Shelf-life and Staling of Chapatti

Introduction

Chapattis are generally prepared fresh and consumed as soon as possible. Freshly baked chapattis are soft, pliable and flexible but as the holding period under ambient conditions progresses, they become leathery, hard and brittle losing their eating quality and consumer acceptability. Texture and flavour are the most important parameters of chapatti quality (Austin and Ram, 1971). The consumer generally judges the texture in terms of softness, flexibility, slight chewiness and ability to be folded into a scoop to hold the curried preparations used as accompaniment. The flavour is judged mainly in terms of sweetish taste and fresh typical wheatish aroma. Storage of chapattis beyond 1 to 2 days under ambient conditions leads to development of off-flavour and mold growth. This decline in freshness of chapatti after baking is attributed to the physical and chemical changes that ensue immediately after its removal from the *tawa* or oven. The changes in physical and chemical properties during storage (aging) are largely attributable to changes in the starch component of starch containing food products (Karim et al, 2000). These time dependent changes may directly contribute or correlate with the changes in sensory perception of the food product leading to staling.

Staling of Baked Products

Staling is a general term that describes time dependent loss in flavour and textural quality of baked products after ceasing of the baking process, and subsequent cooling and storage. Bechtel et al (1953) defined staling as "a term that indicates decreasing consumer acceptance of bakery products caused by changes in crumb, other than those resulting from the action of spoilage organisms". Briefly, staling refers to all changes that occur in bakery products after baking. The most important causes responsible for staling of bakery products are moisture redistribution, starch transformation, and starch gluten interactions (Fadda et al, 2014). During baking starch in the bakery product is gelatinized resulting to an amorphous structure. Upon storage, starch recovers its crystallinity. This process of recrystallization is commonly known as retrogradation, which affects the firming of baked products. Wong et al (2015) have defined retrogradation ".as an ongoing process, which initially involves rapid recrystallization of amylose molecules followed by a slow recrystallization of amylopectin molecules." It is believed that retrogradation is directly related to staling or aging of bakery products (Kulp and Ponte, 1981).The terms staling and retrogradation are often used interchangeably. However, staling refers to all the changes that take place in a bakery product, whereas retrogradation refers specifically to changes in the starch components. Since starch is believed to be the major component responsible for staling of bakery products, most of the methods used to measure staling are dependent on changes in the starch. The methods generally used include sensory evaluation using taste panel, instrumental texture measurement, and changes in starch comprising pasting properties, solubility of starch, enzyme susceptibility of starch, changes in starch crystallinity, changes in X-ray diffraction pattern, changes in differential scanning caloimetry (DSC) to measure changes in enthalpy, etc. (Karim et al, 2000).

Chapatti as compared to bread stales much faster becoming unacceptable within 8-10 h (Sidhu et al, 1990b). The most important accepted indicator of chapatti staling is its texture, which is the attribute most widely recognized by the consumer. The initial stiffening of chapatti within a few hours of baking is similar to firming of pan baked bread crumbs, which is perhaps caused by initial retrogradation of amylose molecules set free during baking (Gray and Bemiller, 2003). Gujral et al (2008) have reported that all the starch present in the flat breads gelatinizes completely during baking even though the baking time is very short. However, there may be incomplete gelatinization of starch granules in the crust of chapatti because of the limited amount of moisture available as reported by Sidhu et al (1997) for Arabic flat bread. This may be one of the factors responsible for rapid staling of chapatti and other flat breads because of the large crust surface area. As the storage period progresses, the texture of chapatti continues to harden, passing from leatheriness to brittleness, which may be the result of progressive changes in the physical structure of amylopectin (Shaikh et al,2007) from a swollen, gel-like state to a more crystalline state. Protein being the third largest component of chapatti dough it is expected to play an important role in staling of chapatti during storage as it has been shown (Srivastava et al, 1998) that the quality of protein contributes greatly to the soft and pliable texture of chapatti, a property most appreciated by the chapatti consumer. Using fractionation and constituent interchange technique between good and poor quality chapatti making whole wheat flours, these authors reported that gluten played a deciding role in influencing the quality of chapatti. It is likely that the protein content and quality also play a role in staling of chapatti as hypothesized in the case of bread staling (Gray and Bemiller, 2003). Gluten-starch interactions and gluten acting as reservoir for moisture transfer to retrograded starch are conjectured to explaining the staling of bakery products including chapatti. Native lipids in particular the polar lipids in whole wheat flour are reported to play a role in staling as these were found to retain soft texture of chapatti during storage (Arya et al, 1978).

Although chapatti has been the staple diet of millions in the Indian subcontinent, it has received scant attention to tackle the problem of staling unlike bread staling which has been studied for more than a century and a half (Gray and Bemiller, 2003). Recent demand for ready to eat chapatti and related products has, however, accelerated the large scale production through mechanization as well as developing ways and means to extend the shelf life of chapatti by curtailing staling. A few studies have been conducted to assess changes taking place during storage of chapattis by testing different staling parameters such as moisture, texture, water soluble starch or water soluble index and water absorption index, in vitro enzyme digestibility, enthalpy change, microstructure, molecular organization, pasting properties, and sensory quality (Parihar and Chatterji, 1956; Nath et al, 1957; Rao, 1982; Venkateswara Rao et al, 1986; Nanjappa et al, 1999, Kaur and Bajwa, 2000; Shaikh et al, 2007 and 2008; Gujral et al,2008; Ghodke, 2009, Hemalatha et al, 2010 and 2014).

Changes in Moisture: Freshly baked chapattis may contain moisture in the range of 28 to 30 per cent (Rao, 1993). However, after cessation of baking chapatti continues to lose moisture and begins to stale resulting in hard texture. There appears to be an inverse relationship between moisture content and staling rate as breads containing higher levels of moisture generally stale more slowly (Kulp and Ponte, 1981; Rogers et al, 1988). The fresh chapatti crust is relatively drier than the crumb resulting in migration of moisture from interior to exterior during storage under ambient conditions till equilibration. The moisture redistribution from crumb to crust perhaps plays a significant role in chapatti staling as indicated for bread (Baik and Chinachoti, 2000). As in bread crumb, moisture transfer perhaps occurs between chapatti components specifically between gluten and starch as chapatti stales. Since water molecules are incorporated into the starch crystallites, the distribution of water is shifted from gluten to starch / amylopectin, thereby changing the nature of gluten network causing further hardening of chapatti. Several studies have been conducted on the loss of moisture in chapattis stored under different conditions. Rao (1982) reported that chapattis kept in the open under ambient conditions lost moisture rapidly and equilibrated to a loss of about 6.8 per cent moisture within 4 h of storage, while those stored packed in waxed paper for 3 days lost about 7 per cent moisture. On the other hand chapattis stored in sealed polyethylene bag or closed plastic box incurred negligible loss in the moisture. This showed that the chapattis kept in the open staled more rapidly than those kept in waxed paper and sealed polyethylene bag or closed plastic box. Venkateswara Rao et al (1986) reported decrease in moisture content from 30.8 to 14.8 per cent in chapattis kept unwrapped in an open container under ambient conditions (27^0C, 66 per cent RH) for 4 h and then reaching an equilibrium value of 5.8 per cent after 24 h storage. On the other hand, there was negligible loss of moisture in chapatti stored either in a sealed polyethylene pouch or plastic container for three days but those wrapped in waxed paper gradually lost moisture probably because of its high permeability to moisture. The chapattis stored in sealed polyethylene pouch or plastic container staled slowly than those kept wrapped in waxed paper and those unwrapped. Kaur and Bajwa (2000) reported a loss of 7.62 per cent moisture in chapattis packed in sealed polyethylene bag stored at 20^0C and 55 per cent RH for three days and remained acceptable. Nanjappa et al (1999) reported 25 to 38 per cent loss of moisture in chapatti and phulka containing chemical preservatives and with or without anti staling agents after storage in paper-aluminum foil-polyethylene laminate pouches for 10 days at room temperature. Lowest loss of moisture was found in phulka containing polyethylene glycol as antistaling agent suggesting that the staling rate is faster in the control phulka and chapatti compared to phulka containing anti staling additives. According to Shaikh et al (2007) the moisture content of chapattis containing preservatives and fat as additives decreased by about 50 per cent after one month storage at room temperature, however, the loss of moisture was comparatively less when stored at refrigerated temperature. On the other hand, Shaikh et al (2008) while studying the inhibition of staling in chapattis by using selected antistaling agents reported that the loss of moisture in chapattis containing antistaling agents during storage for 10 days ranged between about 14 to 24 and 11 to 24 per cent at room temperature and refrigerated temperature, respectively. On the other hand, the chapattis without antistaling agents (control) lost moisture to the extent of about 33 and 26 per cent respectively when stored under similar conditions. Among the additives used guar gum and maltodextrin were found to be more effective in retaining more moisture in chapattis and hence the freshness. Ghodke (2009) studied the effect of guar gum on staling of chapattis by incorporating the gum at 0.25 to 1.0 per cent of whole wheat flour. The moisture content of chapattis decreased during three days of storage at room and refrigerated temperatures. The per cent decrease in moisture content was lowest in chapattis containing guar gum. The chapattis containing 0.75 per cent guar gum experienced the lowest decrease in moisture both at room and refrigerated temperatures, consequently remained much softer than the control chapattis during the period of study. Also, the decrease in moisture contents in chapattis prepared from enzyme treated dough during storage for 4 days at room temperature was low as compared to the control chapattis (Hemalatha et al, 2014). Among the enzymes used bacterial α-amylase and xylanase and their combinations were found to be more effective in retaining the moisture in chapattis indicating that chapattis prepared from dough treated with enzymes retained the freshness better and staled the least. Jha et al (2010) while investigating the rheological and baking properties of different crossed lines of wheat observed a significant negative correlation with dough development time, bread water absorption, bread loaf volume, bread quality score, chapatti water absorption and fresh chapatti moisture content, which indicated that stronger wheat chapattis loose moisture quickly during storage resulting in poor shelf life.

Changes in Physical Properties

Pliability, tearing resistance and shear value are the important characteristics that determine the texture of chapatti. Higher the pliability value, and lesser the tearing resistance and shear force better the chapatti quality. Good quality chapatti should give minimum resistance to tearing (Austin and Ram, 1971).

Pliability: Rao (1982) developed a new simple mechanical device to objectively measure the pliability of chapatti. The pliability decreased gradually in chapatti wrapped in wax coated paper, while it increased slightly in chapatti stored either in polyethylene pouch or plastic box. Open storage without packing resulted in pliability value of zero after 4 h. A highly significant correlation was reported between the moisture content in chapatti and pliability. Venkateswara Rao et al (1986) using the same device, reported rapid decrease in pliability of chapatti from 2.3 to 0.5 cm when kept unwrapped for 4 h under ambient conditions followed by toughening with a pliability value of zero after 8 h of storage. However, chapatti wrapped in waxed paper initially showed a slight increase in pliability at the end of 2h storage probably due to equilibration of moisture within the chapatti. Subsequent storage during three days showed gradual loss in pliability value of chapatti with concomitant loss of moisture. Those chapattis packed in sealed polyethylene pouch rather showed a significant increase in pliability from 2.1 to 4.3 cm after 4h of storage with no further change in pliability value as the storage period advanced. This increase in pliability could be attributed to migration and equilibration of moisture from crumb to the dry crust of chapatti. On the contrary, Kaur and Bajwa (2000) using the same device, have reported progressive decline (from 4.00 to 1.30 cm) in pliability of chapatti packed in sealed polyethylene bag and stored for 72 h.

***Tearing Resistance*:** Rao (1982) adapted a tearing resistance tester normally used for paper and paper board to measure the tearing strength of chapatti. The tearing resistance was reported to decrease with increasing storage period when chapatti was stored in wax coated paper or polyethylene bag but it increased in chapatti kept in the open without wrapping. Venkateswara Rao et al (1986) reported considerable increase in the tearing resistance of chapatti (from 72.5 to 120.0 g) when stored in the open conditions for only 4 h. Chapattis stored in waxed paper showed slight increase in tearing resistance during first two days of storage, thereafter a slight decline in tearing resistance was experienced probably due to the development of brittleness. In the case of chapatti stored in a sealed polyethylene pouch, the tearing resistance decreased considerably after one day of storage then measuring as low as 31 g at the end of three days probably as a result of staling. The increase in the tearing resistance value observed in unwrapped chapatti as well as those stored in waxed paper was probably due to the loss of moisture but the change in tearing resistance of chapatti packed in polyethylene pouch could not be attributed to moisture as no change in moisture content was observed. It could be related to recrystallization and retrogradation of starch in chapatti or to chemical changes not related to moisture content occurring during storage.

***Shear Value*:** The Warner Bratzler (WB) shear value reflects the textural characteristics of chapatti with regard to toughness (see Chapter 6). The WB shear value of chapatti increased considerably from 8.9 to 12.1 lb when stored in open for 4 h indicating the toughening of the texture of chapatti. However, a significant decrease in the WB shear value was observed when chapattis were stored in either waxed paper or in a polyethylene pouch, the decrease being greater in chapattis stored in polyethylene (Rao, 1982; Venkateswara Rao et al, 1986). A highly significant correlation was found to exist between WB shear value of chapatti and sensory texture as well as overall quality of chapatti.

Shaikh et al (2007) measured the texture of chapatti using Stevens LFRA Texture Analyzer with needle probe. The texture of chapatti was expressed as the load in grams required penetrating the product. The fresh chapattis had the least value of 225.26 g indicating that it was soft, whereas the chapattis stored for one month at room temperature showed a value of 618.20 g indicating hardness of chapattis. However, the texture values for chapattis stored at refrigerated temperature were much less than those of chapattis stored at room temperature. The use of emulsifiers/surfactants (Shaikh et al, 2008) as antistaling agents in chapattis, however, improved the texture of chapattis during storage. A combination of sodium stearoyl- lactilate + α-amylase gave the best texture values for stored chapattis indicating a synergistic effect. Hemalatha et al (2010) measured the shear force of chapatti using a texture analyzer fitted with Warner Bratzler blade and reported that the shear force values of chapattis treated with bacterial α-amylase, fungal α-amylase, xylanase, and combination of bacterial á-amylase and xylanase were lower and significantly different from that of their corresponding controls. The lower shear force values denote that the chapattis have soft texture hence easy to tear and fold. Hemalatha et al (2014) further studied the effect of storage on similarly enzymes treated chapattis and observed no significant change in shear force values up to 4 days storage at room temperature.

X-Ray Diffraction: Parihar and Chatterji (1956) investigated the changes in the molecular organization of starch during baking and storage in chapatti using X-ray diffraction technique. The X-ray pattern of chapatti made from wheat flour shows considerable gelatinization compared to pure starch chapatti under identical conditions. This shows that the presence of gluten helps gelatinization of starch during chapatti making. Much less retrogradation of starch was observed in a chapatti with increased gluten content (1:1 ratio of starch and gluten) than in chapatti made from pure starch or wheat flour. The chapattis made from pure starch and wheat flour were found to stale, suggesting that the presence of gluten retarded staling. According to Banecki (1982) gluten is the deciding factor in the staling of bread. The storage temperature was also found to be an important factor in the staling of chapatti. The retrogradation of starch was greatest at 2^0C and decreased with rise in storage temperature. The effect of glycerides was also found to be an important factor in staling of chapatti. In *poori* (a deep fat fried chapatti) the rate of retrogradation was found to decelerate perhaps because of the formation of a fat-amylose complex which perhaps slows down moisture migration.

Thermal Properties: The starch molecules in pastes or gels are known to re-associate upon aging resulting in changes in phase from amorphous to crystallites which brings about changes in energy in the system. The enthalpy change (ΔH) is a useful measurement of staling and evaluation of shelf life of chapatti and related products. It has been widely used by investigators pursuing bread staling (Gray and Bemiller, 2003). Change in the enthalpy (ΔH) is determined using differential scanning calorimeter (DSC) and is expressed in terms of Cal/g or J/g. In DSC, when a thermal transition takes place the energy absorbed by the sample is made up by increased energy input to the sample to maintain the temperature balance. Since this energy input is precisely equivalent in magnitude to the energy absorbed in the transition, a recording of this balancing energy provides a direct calorimetric measurement of the energy of transition which is then recorded as a peak. The area under the peak is directly proportional to the enthalpy change (ΔH). Its direction indicates whether the thermal event is endothermic or exothermic. In the case of retrograded starch, the value of ΔH gives a quantitative measure of the energy transformation occurring during the melting of the recrystallized amylopectin as well as precise measurements of the transition temperatures (onset T_0, peak T_p, conclusion T_c) of the endothermic event.

Nanjappa et al (1999) studied the thermal analysis of retrogradation of phulka and chapatti using the DSC fitted with a thermal analyst system. The thermograms of different fresh phulka preparations showed staling endotherm peak (SEP) at 110^0C which may be due to the retrograded amylose fraction and this occurs immediately in phulka on being cooled to room temperature. This SEP at 110^0C was found to persist in all the stored phulka preparations and is time independent in all the preparations, whereas the SEP temperature due to time dependent retrogradation of amylopectin occurs at about 60 to 65^0C. The SEP for phulka preparations stored for 10 days at room temperature was initially at 60 to 65^0C and then shifted to 80 to 110^0C due to aging. In case of phulka containing margarine and the control, the SEP due to amylopectin merged with that of amylose. The least shift in SEP was observed in phulka containing propylene glycol. Shift in SEP or increase in ΔH value is the time dependent recrystallization of amylopectin which is reported to be well correlated with the rate and extent of retrogradation or staling during storage. The ΔH value of all the phulka preparations increased on storage, however, the extent of increase in ΔH value and shift in SEP differed depending on the type of antistaling additives used. Phulka containing propylene glycol was found to show least rate of retrogradation as indicated by smallest increase in ΔH value. Shaikh et al (2007) measured the ΔH values of chapattis stored at room and refrigerated temperatures during storage period of one month and reported increase in ΔH values with increase in storage period. The thermograms showed a staling endotherm peak (SEP) at about 100 to 120^0C which is due to the retrograded amylose fraction and it takes place immediately in chapatti on being cooled to room temperature. The initial stages of retrogradation appear to be dominated by amylose, however, as the storage period advances further increase in retrogradation is attributed to amylopectin. At refrigerated temperature, the ΔH value decreased initially (up to 5 days) and thereafter it started increasing. This may be because during storage at low temperatures gelatinized starch molecules re-associate, but in less ordered and hence less stable forms than those existing in the native state. Therefore, less energy is required to melt these crystals. Among different antistaling agents (Shaikh et al, 2008) used in the chapattis, sorbitol and maltodextrin resulted in decrease in ΔH values both at room and refrigerated temperature during 10 days storage period as compared to the controls. This shows that retrogradation of starch decreases with the addition of antistaling agents in the dough and there is consequent decrease in the rate of staling of chapattis during storage. Gujral et al (2008) reported that the chapattis, both conventionally (fully) and partially baked, stored under ambient conditions showed higher retrogradation enthalpy than their counterparts kept under frozen (-18^{0}C) conditions for 24 h. This indicates that more energy was required for melting of the starch crystallites in chapattis stored for 24 h under ambient conditions. Also, the fully baked chapattis with lower moisture content (36.7 per cent) than the partially baked chapattis (43.3 per cent) gave significantly lower retrogradation values as compared with partially baked chapattis stored under ambient conditions for 24 h. Retrogradation is very sensitive to the water content in starch gels as recrystallization increases with increasing water content (Zeleznak and Hoseney, 1986). Therefore, the moisture differences appear to affect starch retrogradation in chapattis as well. Sharma and Gujral (2014b) investigated the thermal properties of wheat flour chapattis containing barley flour and β-glucan at different levels, and the effect of storage using DSC. A progressive increase in the enthalpy of gelatinization (ΔH_{gel}) was observed; the increase being more in the case of barley flour addition. Correspondingly, the retrogradation of chapatti during storage at 4^{0}C for 24 h was retarded more by the added barley flour than the added equivalent amount of isolated β-glucan. The studies clearly showed that barley flour and β-glucan both delay staling of the chapattis. However, barley flour was found to be more effective in delaying the staling of fresh chapatti compared to β-glucan at the same level. The dietary fibres (DF) are reported to have greater water holding capacity, therefore addition of barley flour to wheat flour increased the levels of both soluble and insoluble DF which retained more water and also perhaps changed the nature of the starch and protein matrix hence prevented the staling of chapatti more effectively compared to β-glucan singly (Purhagen et al, 2012).

***Pasting Properties*:** Starch, being the major component of wheat flour, plays an important role in the baking and post-baking of baked products. During the course of baking of wheat flour products, starch undergoes the processes of gelatinization, pasting and gelation followed by retrogradation upon storage. Pasting refers to changes in viscosity before, during and after the event of gelatinization, and retrogradation of starch. Maximum paste viscosity or peak viscosity is the viscosity where the starch granules reach the highest swelling while cooking (Yasunaga et al, 1968). Hemalatha et al (2014) used Brabender Micro Visco Amylograph to study the changes in viscosity of the gelatinized starch in the freshly baked chapattis, and the role of added enzymes on the changes in paste viscosity during storage. The control chapattis stored for zero hour showed maximum paste viscosity of 382 BU and cold paste viscosity of 621 BU, whereas those control chapattis stored for 96 h had maximum paste viscosity of 63 BU and cold paste viscosity of 176 BU. Correspondingly, the setback was more for chapattis stored for 96 h showing that stored chapattis had undergone retrogradation. The decrease in paste viscosity of chapattis prepared from the enzymes (fungal α-amylase, bacterial α-amylase, xylanase, and combination of bacterial α-amylase plus xylanase) treated whole wheat flour dough was found to be lower than that of control chapattis; lowest values being for bacterial α-amylase treatment. The ratios of decrease in paste viscosities of chapattis prepared from the fungal α-amylase, and xylanase treated dough were each approximately seven times lesser than that of control chapattis, whereas the ratio of decrease in paste viscosities of chapattis prepared from combination of bacterial α-amylase plus xylanase treated dough was approximately fifteen times less than that of control chapattis. The bacterial α-amylase was found to be more effective in lowering the paste viscosity of chapattis because being heat stable was active in the hot paste as well, thus was found to retard staling of chapattis better than fungal α-amylase and xylanase. Overall, these studies indicated that incorporation of amylases and xylanase in whole wheat flour dough were effective in controlling the staling of chapattis by bringing about suitable changes in the starch and rheological properties of the chapattis. Wani et al (2016) have reported that wheat-pulse composite flours had significantly lower setback viscosity as compared with the control wheat flour; suggesting that composite flours flat breads (chapatti) can maintain freshness for longer period than the control flat bread.

***Changes in Microstructure*:** Hemalatha et al (2014) studied the scanning electron microscopy of both fresh and stored chapattis. The microstructure of chapattis prepared from enzyme treated dough were uniform with distorted starch granules surrounded with protein matrix, while in control chapattis, the starch granules were spherical with protein matrix overlapping on one another to form aggregation. On storage, the microstructure of control chapattis was dominated by protein masses with decreased gelatinized starch granules. On the other hand, there was uniformity in starch structure with in stored chapattis prepared from enzymes treated dough. These differences in structural pattern of the microstructure of control and enzyme treated chapattis could be responsible for changes in texture and mouthfeel of stored chapattis.

Changes in Chemical Properties

During storage the gelatinized starch in chapatti reverts to its native crystalline state and the molecules associate forming crystallites resulting in changes in starch characteristics in respect of lowered susceptibility to enzyme activity, decrease in soluble starch, and decrease in amylose and amylopectin contents, decrease in swelling power (water retention capacity), decrease in damaged starch and decrease in the falling number (Karim et al, 2000).

Nath et al (1957) have shown that both total soluble solids and soluble polysaccharides in chapattis decreased whereas reducing sugars increased during first two days of storage.Venkateswara Rao, et al (1986) examined chapattis stored in polyethylene pouches for changes in the chemical properties during storage.There was no change in the ether extractives, however, total reducing sugars changed gradually and damaged starch decreased drastically after three days of storage. These observations indicated a decrease in enzyme susceptibility of starch. The decrease in water retention capacity and falling number values of stored chapatti further supported the changes in starch characteristics of chapatti during storage.Among different starch fractions, soluble amylose decreased markedly while amylopectin decreased gradually as the storage period advanced suggesting that both soluble amylose and amylopectin are involved in the development of brittleness in chapatti. Kaur and Bajwa (2000) also reported a significant decrease in the hydration capacity as well as soluble solids in chapatti stored in polyethylene pouch for three days.

***Water Soluble Starch*:** Nanjappa et al (1999) reported gradual decrease in water soluble starch (WSS) in phulka and chapatti preparations as the storage period advanced. The rate of decrease in the amount of WSS was faster in control and the chapatti compared to phulka containing different antistaling additives during 10 days of storage. Shaikh et al (2007) observed a significant decrease in WSS in chapattis after one month storage at room and refrigerated temperature. Reduction in water soluble starch in stored chapattis is believed to be owing to association of starch molecules when in paste or gel form resulting in formation of crystallites during storage. These crystallites are relatively less soluble in water than the native starch gel. As the chapatti ages, the amount of water soluble starch extractable from chapatti decreases. When chapatti was incorporated with different emulsifiers/ surfactants (e.g. GMS, SSL, PG, sorbitol, maltodextrin, guar gum, and enzymes) the initial WSS value dropped to between 2 and 3.3 per cent as compared with that of the control (5.23 per cent) at both room and refrigerated temperatures. On the 10th day of storage, WSS values of additives treated chapattis ranged between 0.41 and 1.56 per cent at room temperature, and between 0.56 and 1.85 per cent at refrigerated temperature (Shaikh et al, 2008). According to Ghodke (2009), chapattis incorporated with guar gum (0.75 per cent) showed a decrease in water soluble starch content from an initial value of 29.90 per cent to a value of 6.25, 13.20, and 24.13 per cent on third day at room, refrigerated, and blast freeze temperatures, respectively as compared with the control chapattis with decrease in WSS values of 3.18, 5.46 and 12.45 per cent under identical storage period and temperatures, respectively. These differences in WSS values are reported to be because some surfactants become chemically attached to starch while some mechanically cover starch granules thus reducing their ability to absorb water during mixing. Hemalatha et al (2014) have reported that on storage for 4 days, the soluble starch contents in control chapattis decreased by 72 per cent, however, chapattis prepared from dough treated with bacterial α-amylase, and fungal α-amylase registered decreases by only 2 and 6 per cent, respectively. The soluble starch contents of chapattis treated with xylanase weresimilar to that of control chapattis; however, storage (4 days) resulted in decreases, being lesser in treated chapattis compared to the control. Combination of bacterial α-amylase and xylanase did not change soluble starch content of chapattis during first 2 days of storage but on 4th day of storage it decreased by 5 per cent. The studies indicated that the incorporation of these enzymes in dough retard staling of chapatti by bringing about changes in the starch and rheological properties of chapattis.

***Water Absorption and Water Solubility Index*:** Gujral et al (2008) determined the water absorption index (WAI) and water solubility index (WSI) to study the extension of shelf life of conventionally (fully) and partially baked chapattis under ambient and frozen (-18^{0}C) storage conditions. The starch molecules upon gelatinization during chapatti baking start associating during storage forming crystallites that exhibit reduced swelling capacity or water absorption index, and less solubility in water or water solubility index. The results showed decline in WAI of conventionally baked chapattis when stored for 24 h at both ambient and frozen temperatures, percentage decrease being lower in the case of chapattis stored under frozen conditions. This may be attributed to the formation of amylopectin crystallites with ordered structure. However, the retrogradation of starch was minor under frozen conditions. The WAI of partially baked chapattis, on the other hand, incurred a higher percentage decrease under ambient conditions but much lower percentage decrease under frozen conditions during 24 h storage. The higher decrease in the WAI may be attributed to the fact that the retrogradation was higher when the chapatti had higher moisture content and this correlates well with the higher retrogradation enthalpy observed in partially baked chapatti stored under ambient conditions. The lowest percentage decrease in WAI of chapattis stored under frozen conditions is indicative of less crystallite formation, in other words lower retrogradation under frozen storage. The water soluble index (WSI) of the conventionally (fully) baked chapattis showed non-significant increase during storage under ambient conditions but showed significant increase under frozen conditions. On the other hand, the WSI of partially baked chapattis showed significant decrease when stored under ambient conditions but there was increase in WSI upon storage under frozen conditions. The increase in WSI observed for both fully and partially baked frozen chapattis may be attributed to the ice crystal damage that occurs during the freezing process. The large ice crystals damage and rupture the chapatti structure and result in the formation of soluble fragments thus increasing the WSI. The maximum decrease in the WSI observed in the partially baked chapatti kept under ambient conditions for 24 h was suggestive of the maximum retrogradation suffered by it.

In vitro Enzyme Digestibility: *Invitro* enzyme digestibility (IVED) of starch is determined by digesting the food product with amyloglucosidase enzyme. This enzyme hydrolyses the amorphous portion of the starch but is chemically inert towards crystallites starch. The formation of crystallites as a result of retrogradation increases the resistance of starch to enzymatic hydrolysis. The resistance of starch to enzymatic hydrolysis can be used as a measure of starch retrogradation. As the per cent crystallinity of starch increases, the IVED decreases. Nanjappa et al (1999) reported progressive decrease in the IVED of phulka preparations with increase in storage time. However, there was comparatively less decrease in IVED in phulka containing propylene glycol, and other additives. Resistant starch (RS) is formed during storage of chapatti and phulka. The amount of RS depends upon the percentage of crystallinity. Maximum amount of RS was found in control and chapatti and least in phulka containing propylene glycol. IVED is inversely proportional to RS; hence phulka containing propylene glycol stales comparatively less and slowly. Similar observations were made by Shaikh et al (2007) who reported a decrease by about 55 per cent in IVED of chapattis during one month storage at room temperature. This is because of continued recrystalization of amylose and amylopectin during storage leaving minimum amount of amorphous amylose and amylopectin for the enzyme amyloglucosidase to act upon. When treated with different surfactants/emulsifiers, the IVED of chapattis during 10 days of storage decreased between about16 to 37 per cent at room temperature and between about 14 to 28 per cent at refrigerated temperature in comparison with control chapattis which had IVED per cent reduction of 34 at room temperature and 30 at refrigerated temperature for the same period of storage (Shaikh et al, 2008). Alpha amylase was found to be more effective in controlling the crystallization of the starch in chapatti as it showed minimal reduction in IVED values both at room and refrigerated temperatures. Ghodke (2009) while studying the effect of guar gum on staling in chapatti also reported decreases in IVED values from 2.30 to 1.84, 1.21, and 1.01 at room temperature during storage for three days. This shows that during storage limited amounts of amylose and amylopectin due to recrystallization, are available for the action of the enzyme resulting in lower values for IVED. On the other hand, at refrigerated and blast freeze temperature storage, the IVED values are higher may be because during storage at low temperatures gelatinized starch molecules reassociate, but in less ordered and hence less stable forms than those existing in the native state. Addition of guar gum affects amylose network avoiding formation of spongy matrix. The hydrophilic character of the gum prevents water release and polymer aggregation during refrigeration.

Changes in Sensory Quality

The extent of changes in sensory properties of chapatti during storage under ambient conditions was found to depend upon the type of packaging material used.Rao (1982) observed loss of typical wheaty aroma and sweetish taste in chapattis during storage. The chapatti stored for three days in wax coated paper was found bland in taste and flavour, while that stored in polyethylene bag or plastic box developed slight off flavour. The stale aroma in stored chapattis possibly develops due to the changes in volatile carbonyl compounds produced during baking of chapatti. There is increase in undesirable α, β-unsaturated carbonyls via aldol condensation reaction during storage (Arya, 1984). Venkateswara Rao et al (1986) showed that chapatti stored unwrapped became hard, gritty and brittle as well as dry after 4 h of storage under ambient conditions. The chapattis packed in waxed paper developed a bland taste and flavour with deterioration in eating quality at the end of 3 days of storage. The chapattis packed in sealed polyethylene pouch became slightly tough and brittle after 24 h of storage followed by development of mild off-flavour at the end of three days of storage. Kaur and Bajwa (2000) also observed a significant decrease in overall acceptability scores of chapattis stored in sealed polyethylene bag after two days of storage.Shaikh et al (2007) observed a decrease in sensory quality and acceptability of chapattis stored in a self-sealing low density polyethylene(60 gauge) plastic bag at room temperature for one month, whereas at refrigerated temperature the sensory scores decreased with time of storage reflecting loss in the acceptability of chapattis. On the other hand, the sensory scores for chapattis treated with different antistaling agents (Shaikh et al, 2008) decreased less as compared with the controls during storage for 10 days at both room and refrigerated temperatures. However, the chapattis treated with maltodextrin and SSL+ α-amylase combination received higher sensory scores as compared with the other antistaling agents used in this study. Hemalatha et al (2014) have shown that unlike control chapattis, the overall sensory quality scores of chapattis prepared from dough treated with enzymes (amylases and xylanase) did not change significantly upon storage in sealed polyethylene pouches for 2 days at room temperature. Chapattis prepared from dough treated with bacterial α-amylase, however, received comparatively higher sensory quality scores than the chapattis prepared from the dough treated with other enzyme systems. The improvements in chapatti quality observed upon addition of amylases and xylanase may be due to changes in starch and pentosan properties, respectively. In addition to improving the sensory quality and shelf-life of chapattis, treatment with amylases and xylanase had enhanced the nutraceutical andantioxidant properties of chapattis as well.

Inhibition of Staling and Extension of Shelf-life of Chapatti

Freshly prepared chapattis are highly perishable and have a shelf-life of 1 to 2 days after which they develop off-flavour and mold growth. Therefore, extending the shelf-life of chapatti has always been sought after.

Chapatti Storage at Household Level

Chapatti-making is a cumbersome process involving kneading of dough, rolling the dough balls into thin, circular dough discs and then baking on a hot *tawa* followed by puffing on the open flame or on the hot *tawa* itself by gentle pressing with a cloth pad. If one eats chapattis every day, one needs not to take the trouble of going through the above cumbersome process. Use of suitable methods to extend the shelf life of fully or partially baked chapattis may be practiced.

Preserving the chapatti for a week or a month requires that it should be soft in texture with a pleasant sweet taste and aroma. For making soft chapattis, use of warm milk or *dahi* (yogurt) is advocated instead of water when kneading the dough. The dough should be kneaded well at least for 15 min, and kept aside covered with a wet cloth for 30 to 60 min before rolling and baking the chapattis. The baked chapattis should be cooled to room temperature and then packed in polyethylene bag or aluminum foil or in a Ziploc bag or a freezer bag to prevent ingress or loss of moisture. Thus packed chapattis are then placed in the freezer. Alternatively, it would be better to freeze and store the rolled chapatti dough, i.e. the raw chapatti, defrost them and then bake them in the usual way, which would be as good as fresh chapattis. Before reuse, remove the fully baked frozen chapattis from the freezer and defrost them for 1 to 2 h while in the bag. The defrosted chapattis may be heated on a preheated *tawa* or in a microwave oven. For convenience, do not store all the baked chapattis together in one bag for freezing, store in portions so that you can use each portion according to your needs. Place a piece of wax paper in between the chapattis to prevent sticking. Also, the defrosting of frozen chapattis should be preferentially exercised by placing them in the refrigerator to curtail contamination.

For storing chapattis for 2 to 3 days, wrap a stack of required number of baked and cooled chapattis in aluminum foil and place the foil stack in a sealable polyethylene bag and store in a refrigerator. Before serving, heat the chapattis on a preheated *tawa* or in a microwave oven or a conventional oven.

Chapatti Storage at Commercial Level

The rapidly changing lifestyles, urbanization and industrialization with increasing number of women working, has created increasing demand for convenience foods in particular the ready-to-eat and ready-to-cook food products. Chapatti is no exception in this regard and is receiving urgent attention in view of the recent developments in mechanization of industrial level chapatti production (Sridhar and Manohar, 2001). The chapattis produced on large scale will require marketing in suitable unit packs. The packed chapattis should retain all the characteristics of fresh chapattis as referred to above and remain microbiologically safe. This will require a double prong approach, firstly controlling the process of staling to retain the soft, pliable texture of chapatti, and secondly controlling the microbes to provide not only reasonable shelf life but also ensuring consumer acceptability. This could be accomplished by, manipulating the processing techniques (see Chapter 4), use of suitable preservatives / additives, improved packaging techniques and storage conditions, irradiation, freezing, and thermal processing. The developments that have taken place in this direction are discussed below:

Use of Preservatives/Additives

Since the preparation of chapatti does not generally involve the use of common salt and or sugar like in the case of pan bread, the water activity of chapatti is slightly higher than that of the pan bread of equal moisture content. Common salt and sugar are known to lower the water activity of a food product. The freshly baked chapattis may contain moisture in the range of 28 to 30 per cent with ERH between 88 to 92 per cent, and pH value close to 5.5 (Arya et al, 1977). These factors coupled with the tropical and semitropical conditions prevailing in the Indian subcontinent encourage the growth of mold and spoilage of chapatti. Moreover, the recent advent of automation in the production and packaging of chapattis in the Indian subcontinent has extended the time interval between baking and consumption. Consequently, the need has arisen to extend the shelf-life of chapatti- the staple food of the millions.

The effect of preservatives on the keeping quality of chapatti has been investigated by several workers. Initial attempts by Nath et al (1957) to arrest the mold growth by the addition of propionates in chapattis were not fully successful. However, Yasin et al (1965) showed that the addition of 0.15 per cent propionate or sorbate to chapatti formula was found sufficient to retard mold growth for up to five days. Rao et al, (1964; 1966) reported that chapatti containing salt (2.0 per cent) and sorbic acid (0.48 per cent) on wheat flour basis, packaged and heat sealed in paper-foil-polyethylene pouch, stored satisfactorily and remained acceptable for six months under diverse tropical conditions. During storage fat acidity increased slightly, the reducing sugars increased whereas non reducing sugars decreased. The addition of hydrogenated fat was found to retard the retrogradation of starch and improve the texture of the preserved chapattis. But, the stored preserved chapattis did not receive wide acceptance because of development of slight bitter aftertaste. Hussain and Satti (1969) too found that sorbic acid imparted a bitter aftertaste to *roti* and, therefore, was not recommended as a preservative for it. Moreover, the concentration of sorbic acid in chapattis (0.25 per cent as is basis) was higher than the level (0.15 per cent) permitted by the Indian Prevention of Food Adulteration Act, 1954 (Now: Food safety and Standards Act, 2006). This necessitated reducing the concentration of sorbic acid to improve the flavour and texture of preserved chapattis. Arya et al (1977) were instrumental in reducing the effective concentration of sorbic acid to 0.20 per cent (wheat flour basis) by supplementing with citric acid (0.4 per cent), salt (2.5 per cent) and sugar (3 per cent) followed by in pack sterilization of chapattis for 2 h at 90^0C. The residual concentration of sorbic acid in the chapattis thus preserved varied from 0.09 to 0.10 per cent on as is basis which is less than the permitted level. Both citric acid and sugar are known to suppress the dissociation of sorbic acid in aqueous solutions. Since only undissociated sorblc acid has antifungal action, inclusion of citric acid and sugar in chapatti formula proved highly beneficial in preventing mould spoilage with low levels of sorbic acid. Both salt and sugar are reported to increase the efficiency of sorbic acid against molds. Sugar also helps to mask the slight bitter taste that develops during storage in preserved chapattis. Moreover, added salt and sugar will lower the water activity of chapatti thus further increasing the shelf life. Incorporation of fat, milk powder and oleoresins of cardamom and cumin seeds were found to improve the texture and flavour of preserved chapattis. Vijaya Rao et al (1979) carried out detailed microbial studies on in-pack heat processed chapattis preserved using similar combination of preservatives as used by Arya et al (1977). They concluded that combination of packaging soon after baking, in-pack heat treatment, low pH and presence of sorbic acid was conducive to good shelf-life of preserved chapattis. The in-pack heat treatment of chapattis eliminated the contamination picked up after baking from handlers, equipment surfaces, and processing premises environment. Vidyasagar and Arya (1981) reported that sorbic acid used for preserving chapattis did not cause any significant changes in the biochemical reactions taking place during dough mixing and baking of chapattis. Rao et al (1974) have earlier reported that the sorbic acid added to chapattis had no deleterious effect either when consumed freshly made or after six months storage. The studies carried out by Rao (1982) on chapatti showed that addition of sodium propionate (0.20 per cent or sorbic acid (0.20 per cent) delayed the onset of mould growth by one day as compared to the control but did not affect the textural and flavour characteristics of chapatti. Addition of sorbic acid (0.48 per cent of wheat flour) had no significant effect on the protein efficiency ratio and feed efficiency ratio of chapattis stored for six months (Prasad et al, 1987).

Ghosh et al (1973; 1977) advocated the use of fungistatic wrappers to preserve chapattis and similar products. Fungistatic wrappers were prepared by coating grease proof paper with the coating formulation comprising sorbic acid, carboxymethyl cellulose, Embanox-6 (antioxidant) and water. The coated paper contained 2.0 g sorbic acid per square meter. The chapattis were first wrapped in the coated paper and then enclosed in laminated paper-foil-polyethylene pouch followed by heat processing at 90-95^{0}C for 30 min. The product had a shelf-life of more than 3 months based on microbiological stability and two months based on organoleptic acceptability. However, the process of heating may be eliminated when the chapattis are packaged immediately after baking. Chapattis remained mold free for more than two months with a useful shelf-life of 30 days. It was also observed that the amount of sorbic acid transferred to food from the wrapper is well within the limit of 0.1 per cent permitted under the Prevention of Food Adulteration Act. Chapattis preserved with sorbic acid or sorbates / propionates are, however, reported to develop slight bitter after taste during storage (Arya, 1984).

Food additives are commonly used not only to improve the sensory quality but also the keeping quality of baked food products. The additives that have been used in chapattis include: non-fat-dry-milk (NFDM). yogurt, butter milk, yeast, emulsifiers or surfactants, e.g. glyceryl monostearate (GMS), sodium steroyl-2- lactilate (SSL), sucrose esters, enzymes (á-amylase, xylanase), hydrocolloids such as guar gum, carboxymethylcellulose (CMC), hydroxypropylmethylcellulose (HPMC), k-carrageenan, and several other additives.

Hussain and Satti (1969) prepared Pakistani *roti* with GMS and added vital gluten in dough, which remained soft and resilient for up to 72 h at 35^{0}C. The use of GMS and NFDM slows down the loss of moisture, making chapattis / *rotis* retain freshness for longer period. Also, the addition of *vanaspati* (hydrogenated fat), and mono-glycerides has been found beneficial in the retention of soft texture in preserved chapattis for longer period (Arya et al, 1977). Swaranjeet et al (1982) reported the feasibility of delaying staling and extending the shelf life of chapattis by leavening and supplementation with GMS for up to 12 h after baking and keeping in air tight container. The chapattis when refreshed by warming on a hot plate were relatively much softer than the control chapattis after 24 h. The addition of shortening (3 per cent) in combination with GMS (0.5 per cent) in the whole wheat flour formulation produced chapattis which were initially soft and flexible, and showed the least amount of firming during 3 days storage (Sidhu et al, 1989). Chapattis prepared from dough acidified with lactic acid and/or yeasted for2 to 4 h showed improvement in colour, texture and shelf life (Syed et al, 1993). Additives like wet gluten, liquid shortening, GMS, CMC, sodium caseinate and skim milk powder were found to significantly increase the extensibility of chapatti and kept the chapatti more extensible even after 24 h of storage at 25^{0}C (Gujral and Pathak, 2002). The enzyme diastase, however, did not significantly affect the extensibility of fresh chapatti but the chapatti remained more extensible than the control chapatti after 24 h storage. Barley flour, wet gluten and liquid shortening in various combinations markedly improved the quality of fresh as well as stored (24 h) chapatti (Gujral and Gaur, 2002).Shaikh et al (2008) while investigating the effect of addition of SSL, GMS, propylene glycol, sorbitol, α-amylase, xylanase, maltodextrin and guar gum to chapatti dough on the storage quality of chapattis at 29 ± 1^0C and 4 ± 1^0C reported that all the additives inhibited staling of chapatti to different extent. However, maltodextrin (0.3 % WW) and combination of α-amylase (100 ppm) and SSL (0.375%) were found best to extend the storage life of chapattis.Khan AN et al (2011) recommended the use of carboxymethyl cellulose, lecithin, sodium propionate and ascorbic acid in whole wheat flour as these improve not only the sensory attributes but also extend the shelf-life of chapattis. Hemalatha et al (2014) have reported that the incorporation of amylases and xylanase in whole wheat flour dough retard staling of chapatti by bringing about changes in the starch and rheological properties of chapattis resulting in extension of shelf life of chapatti.

Several improver premixes / blends have been developed and patented, which when used in dough are reported to not only improve the sensory quality but also extend the shelf life of chapattis. Manohar and Rao (2005) have developed and patented a novel improver premix comprising ascorbic acid, sodium steroyl-2-lactilate and guar gum for improving the quality of chapatti with longer shelf life. Khan et al (2006) developed and patented a blend of additives containing lower level of sorbic acid coupled with nisin – a biopreservative to extend the shelf-life of chapatti for one year without any deleterious effect on consumer acceptability. Foreman and Evanson (2010) patented the use of selective enzyme amylases which are capable to degrade raw starches at low temperature for retarding of staling process during storage thus extending the shelf life of flat breads. The process may be useful for industrial production of flat bread chapatti and similar products.

Irradiation Preservation

Irradiation offers a new technique for the extension of shelf-life of chapattis without having to use chemical additives which affect the organoleptic qualities. Savagaon et al (1970) exposed chapattis, sealed individually in polyethylene bags to CO^{60} gamma rays at a dose rate of 1.2 Mrads per hour. The irradiated chapattis were stored at room temperature (25-30^0C) in covered boxes. The irradiated chapattis remained free from mold growth for more than six months and were highly acceptable even after storage for 60 days though the total sensory score decreased with time perhaps due to staling phenomenon. From the results of acceptability, keeping quality and absence of peroxidation, it could be feasible to extend the shelf-life of chapattis up to 60 days under ambient conditions by a radiation dose of 1 Mrad. It was postulated whether combination of irradiation (at lower dose level) and heat treatment could stabilize the prepackaged chapattis against mold growth during prolonged periods of storage. Padwal Desai et al (1973) subjected chapattis packed in Polycell pouches to 50 Krad irradiation followed by dry heat (65^0C) treatment. The irradiated chapattis remained free from mold and shelf stable for 10 weeks at ambient temperature (28- 32^0C). There were no immediate discernible changes observed in organoleptic attributes of chapattis even after 100 Krad irradiation. Marathe et al (2002) irradiated prepackaged whole wheat flour at doses up to 1 kGY and found no adverse effect on chemical composition, dough properties, total bacterial and mold count up to 6 months storage. However, irradiation at 0.25 kGY was found sufficient to extend the shelf-life of *atta* up to six months without any significant changes in the nutritional and functional attributes. Chapattis prepared from irradiated (0.25 kGY) *atta* were preferred even after 6 months storage compared with the control. Irradiation and subsequent storage, however, result in very pronounced off-odors probably due to the radiolytic degradation of amino acids and fatty acids. Addition of antioxidants has not proved effective in retarding the development of off-odors in irradiated chapattis. But irradiation at very low temperatures (frozen state), which has proved effective in retarding the off-odors in some meat products, may prove beneficial in improving the acceptability of irradiated chapattis (Arya, 1984). Bhoir et al (2015) preserved *methi paratha* by vacuum packaging in multi-layered pouches and irradiating at 25 kGy in frozen conditions. The *parathas* remained microbiologically stable for 180 days.

In view of the increasing consumer awareness and aversion to accept food products preserved using chemical preservatives and irradiation, thermal processing and freeze preservation are perhaps the alternatives to extend the shelf life of chapattis.

Thermal Processing

***In-pack dry-heat processing*:** Venkateswara Rao et al (1981) conducted thermal sterilization of chapattis sealed in 25 and 50µm polyethylene bags at 100^0C for 15- 16 min. The sterilized packages were then stored at 26- 28^0C and 75- 80 per cent RH. Chapattis in the thinner packaging developed mould growth after 10- 15 days, whereas those in 50µm packages were free of mold even after 15 days regardless of sterilization time. The sensory quality of sterilized chapattis in 50µm film was found highly acceptable. Rao (1982) reported considerable delay in the onset of mold growth when chapattis were subjected to in-pack dry heat treatment depending on the duration of sterilization. The chapattis heated for 40-50 min at 90^0C stored well for 6- 9 days without any mold growth with slight improvement in flavour of these chapattis as compared to chapattis stored without any heat treatment. A longer period of 60 min or more heat treatment of chapattis though delayed the mold growth considerably; it had significant adverse effect on the texture and flavour of chapattis. However, in-pack heat processing of chapattis containing salt (1.25 per cent), fat ((4.0 per cent) and glyceryl monostearate (0.5 per cent) for a period of 50 min at 90^0C was found to be the best treatment extending the storage for about 12 days without the onset of mold growth and without any significant changes in textural and flavour characteristics of chapattis. Mathur et al (1973) attempted to preserve chapattis by thermal processing in cans. It has been claimed that chapattis could be preserved in acceptable condition for more than a year by processing at 116^{0}C for 100 minutes.However, the quality of thermally preserved chapattis has been found to be far inferior to the one preserved by sorbic acid. This is mainly because of excessive caramelization and the resultant hard and leathery texture of canned chapattis.

Retort Processing

Despite huge marketing potential, the commercial marketing of chapattis preserved by using the methods and techniques discussed above did not pick up well due to high perishability and declining freshness appeal. This, however, has been reversed using the retort processing technology for extending shelf life of chapatti with acceptable sensory appeal, i.e. better colour, texture, taste and mouthfeel.

Retorting is a process that uses heat and pressure to raise the temperature of the food in a strong and sealed package that can withstand required heat and pressure. The packaging may be rigid, semi-rigid or flexible. The flexible packaging used is generally less bulky than can and jar. Food is heated more quickly retaining the original taste and flavour of the product in the sealed packaging. The process uses water or steam/air combination as processing medium to heat the packaging in a retort (autoclave). The ready-to-use retort pouches are flexible packages made from multilayered plastic films with or without aluminum foil as one of the layers. These materials are heat resistant and suitable for processing in a retort at a temperature of around 121^{0}C. The retort pouches are tough and puncture resistant, heat sealable, compatible with food product, and have good barrier properties (Ramaswamy and Marcotte, 2006).

Following reports substantiate the usefulness of retort processing to extend the shelf life of chapattis and related products with retention of desired sensory qualities.

Badgujar et al (2010) retort processed the chapattis sealed in retort pouches at different F_0 values, e.g. 0.5, 1, 2, 3, 4, 5, 6 and 7 min. The sterilized pouches were immediately cooled to 50^{0}C using cold water. The processed chapattis were then stored at accelerated temperature (40^0 C) for 12 months. At regular intervals, the chapattis were evaluated for microbial count, colour, tearing force, extensibility and sensory quality. The results showed that the retort processed chapattis were shelf-stable (zero microbial count) for more than 12 months. However, the tearing force of chapatti which reflects softness or hardness increased from 251.7 to 1396.88 g during 0 to 90 days of storage. The retort processed chapattis were, therefore, found unacceptable in terms of colour, texture and overall sensory acceptability. However, addition of improvers (sodium steroyl-lactilate, glyceryl monostearate and diacetyl tartaric acid ester mono- and di-glycerides) resulted in slight improvement of chapatti colour, texture and sensory quality.

Khan MA et al (2011) have developed convenient, ready-to-consume, thermally processed chapattis having natural sensory attributes. Whole wheat flour (1 kg) of 100 per cent extraction, prepared in the laboratory plus 22g common salt, 50 g hot molten *vanaspati* (hydrogenated fat) and 650 ml water were mixed stepwise in a dough kneader into dough of desired consistency. The dough was rested to condition and mixed again. The dough balls each weighing about 45g were flattened into chapatti dough discs (15 to 17 cm diameter) using a semi-automatic flattening machine. The chapattis were baked on a hot plate maintained at around 220 to 230^{0}C. Molten *vanaspati* was applied to the chapatti during baking. Four chapattis having butter paper lining in between each pair were packed in retortable pouches and sealed immediately. The sealed pouches were placed in a retort for processing. The come-up-time of the product to attain 100^{0}C was 5 to 6 min. After attaining 100^{0}C, the temperature of the product was raised to120^{0}C and maintained for 14 min. Pressure was maintained at 20 lb with air (5lb) and steam (15lb) throughout the process using steam-air during heating and air-water while cooling. After processing the product to required F_0 value of 3.0, the pouches were cooled rapidly till the core temperature reached to 75^{0}C by pumping water into the retort and circulating it. The pouches were wiped dry and stored at ambient temperature (15 to 35 ^{0}C) conditions. The chapattis remained microbiologically stable and highly acceptable even after a storage period of 12 months. **Figure1** represents the flow sheet for the preparation and retort processing of chapattis.

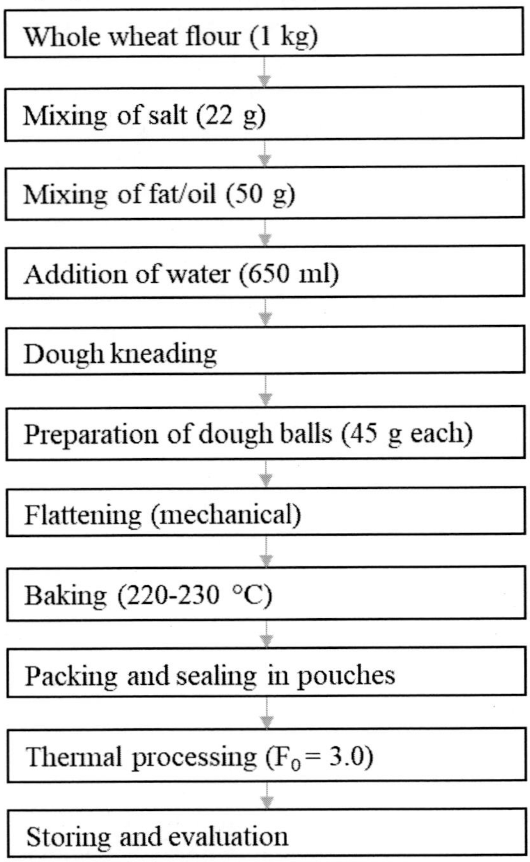

Figure 1: Flow sheet for preparation and retort processing of chapattis (Source: Khan MA et al, 2011)

Kotturi and Kannampilly (2013) prepared chapattis using standardized recipe comprising whole wheat flour (100 g), oil (10 ml), salt (1.8 g) and water (60 ml). The chapatti dough discs were sheeted by pressing mechanically and baked on a hot plate (250^{0}C) for 12 s on one side and 10 s on the other side. The partially baked chapattis packed in polypropylene pouches of 100 microns were retort processed to achieve commercial sterility in a pilot-scale steam-air retort maintained at 95^{0}C. Overhead air pressure of 10 psi was used during processing and cooling. To attain commercial sterility a minimum time of 4.5 min was required. The flow sheet showing preparation and retort processing of partially baked chapattis is presented in **Figure 2**.The partially baked retort processed chapatti pouches stored at ambient temperature remained microbiologically safe for more than eight weeks.

The retort processed partially baked chapattis stored at ambient conditions were rebaked for 15 s on one side and 10 s on the other side. The rebaked cooled chapattis were evaluated organoleptically for appearance, pliability, aroma, texture, and overall eating quality. The chapattis remained highly acceptable even after storage for 8 weeks under ambient conditions. However, the overall quality scores decreased slightly from 9.2 to 8.08 during the storage period of 0 to8 weeks.

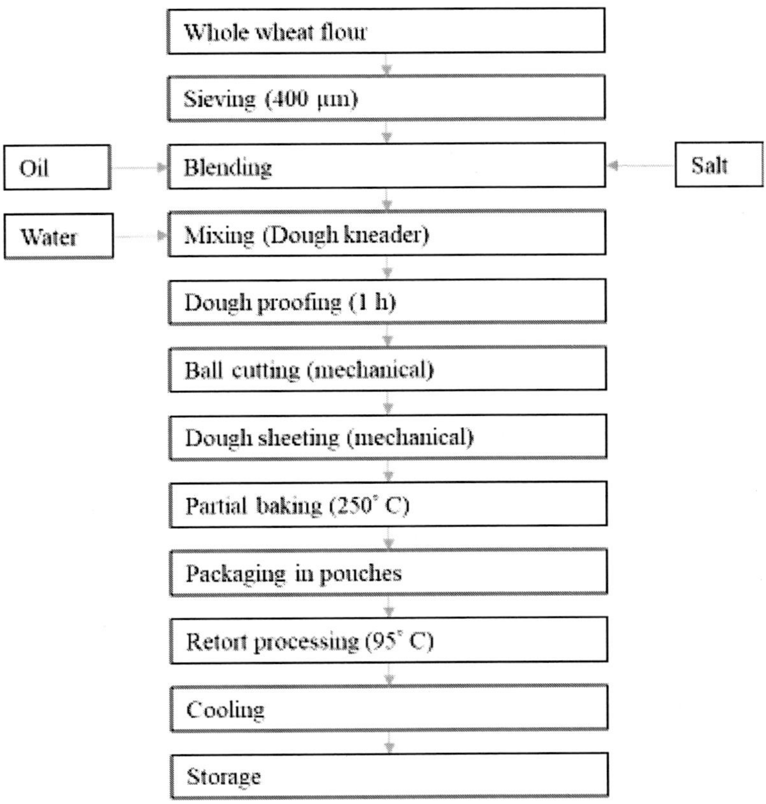

Figure 2: Flow sheet for preparation and retort processing of chapatti (Source: Kotturi and Kannampilly, 2013).

Freezing Preservation

Another approach for extending the shelf life of chapattis may be the application of sub-zero temperatures in which fully baked or partially baked chapattis are frozen, thawed and then used with or without rebaking as applied in the case of bread making (Rosell and Gomez, 2007). Following reports have explored the potential of freezing preservation of fully or partially baked, and ready-to-bake chapattis:

***Freeze Preservation of Fully baked Chapatti*:** Gujral et al 2008) sheeted the whole wheat flour dough into chapatti dough disc manually and baked it on hot plate (250^{0}C) for 15 s on one side and 20 s on the other side and again turned to bake for 20s. After cooling for 10 min at ambient temperature, the chapattis were packed in aluminum foil and polyethylene pouches for storage at ambient temperature (25^{0}C) in an air tight container for 24 h. Another batch of chapattis similarly packaged was kept frozen at -18^{0}C for up to 4 weeks. The fully baked frozen stored chapattis were evaluated for textural qualities after thawing for 2 h at room temperature (25^{0}C).The textural properties of fresh as well as stored chapattis were measured on Instron Universal Testing Machine. The fresh chapattis were soft and extensible as shown by the low values of peak force and modulus of deformation as well as by the longer distance of extension i.e. extensibility. However, the extensibility decreased while the peak force, modulus of deformation and the energy to rupture increased when chapattis were stored for 24 h at room temperature (**Table 1**). On the other hand, the fully baked frozen chapattis stored for 24h at -18^{0}C showed a decline in extensibility. The extensibility value was lower than the fresh control chapatti but was significantly higher than the chapatti stored at room temperature for 24 h. The peak force for fresh and frozen chapatti was not significantly different from each other but the chapatti stored at room temperature for 24 h showed significantly higher peak force values as compared to the fresh control chapatti. The chapattis both fresh and stored frozen at - 18^{0}C for 24 h showed similar values for energy to rupture.

Table 1: Textural qualities of fully and partially baked chapattis stored under different conditions (room temperature or -18^{0}C).

Chapatti	Storage	Extensibility (mm)	Peak force (N)	Modulus of Deformation (MPa)	Energy to Rupture (J)
Fully baked	Fresh	10.37	1.6	0.38	0.008
Partially baked	Fresh	13.23	1.8	0.36	0.011
	24h ambient	12.74	2.8	0.53	0.018
	24h frozen	13.10	2.6	0.51	0.017

Source: Gujral et al (2008).

***Freeze Preservation of Partially baked Chapatti*:** For the preparation of partially baked chapattis, Gujral et al (2008) baked the sheeted dough disc on a hot plate (250^{0}C) for 15 s on one side and 10 s on the other side. The partially baked chapattis were stored both at ambient temperature and frozen temperature (–18^{0}C) as described above for fully baked chapattis. The frozen partially baked chapattis were removed from the freezer and thawed for 2 h and then rebaked to completion at 250^{0}C for 15 s on one side and 10 s on the other side and cooled to room temperature for 10 min before evaluation for textural qualities. For comparison, the texture profiles of partially baked chapattis after rebaking and that of fully baked chapattis are shown in Table 1.The extensibility of the partially baked chapattis after rebaking and stored at room and frozen temperatures was not significantly different from each other but were significantly higher than the extensibility of the fully baked chapattis. The textural qualities of partially and fully baked chapattis on extended frozen storage for 2 and 4 weeks at -18^{0}C differed with the process of baking as well as period of storage. The extensibility of fully baked chapatti decreased initially after 24 h of frozen storage but with no further decrease at 2 and 4 weeks of frozen storage. The extensibility of partially baked chapattis, on the other hand, remained statistically unchanged during the first 24 h of frozen storage but showed a significant gradual increase when stored for 4 weeks in the freezer. The peak force and modulus of deformation of the fully baked chapattis increased up to 2 weeks of frozen storage and then declined at 4 weeks of frozen storage. In the case of partially baked chapattis, the peak force increased at 24 h of frozen storage and then remained statistically unchanged up to 4 weeks of frozen storage, whereas the modulus of deformation increased during the first 24 h of frozen storage with subsequent decrease at 2 and 4 weeks of frozen storage. The energy to rupture the fully baked chapattis was highest at 2 weeks of frozen storage whereas the partially baked chapattis showed an increase in the energy to rupture with increasing frozen storage period up to 4 weeks. These differences in the tensile deformation behavior of the fully and partially baked chapattis during frozen storage could be due to the differences in their initial moisture contents. The studies reveal that the frozen partially baked chapattis after thawing and rebaking exhibited texture equivalent to that of fully baked chapattis, therefore, they could be considered a better option than the frozen conventionally baked chapattis for slowing the process of staling.

Ahmed et al (2013) studied improvement in the quality, storage stability and shelf life of partially baked chapattis by supplementing with different hydrocolloids (0.5 per cent) plus sodium propionate (0.3 per cent) followed by frozen storage for 28 days. The chapattis were prepared from whole wheat flour according to the method described above by Gujral et al (2008). The rheological studies showed an increase in water absorption, dough development time and dough stability after addition of the hydrocolloids. There was increase in moisture retention and water soluble starch. Supplementation with the hydrocolloids improved the baking quality, sensory attributes as well as shelf life as compared to the controls. Among the hydrocolloids hydroxypropylmethyl cellulose showed the most favourable results followed by carboxymethyl cellulose and guar gum. The combination of partial baking, supplementation with hydrocolloids and frozen storage helped improve the shelf life and quality of partially baked chapattis.

***Freeze preservation of ready to bake chapatti*:** The ready-to-bake bakery products have been well accepted in the developed western nations and the concept is gaining importance in other parts of the world as well as it offers convenience, and the product has longer shelf life and can be marketed directly to the consumer. The products made from frozen dough are similar to their freshly baked counterparts and offer savings on equipment and labour (Kenny et al, 1999). Yadav et al (2009b) studied the effect of prolonged frozen storage (-18^{0}C) on the textural and sensory properties of ready-to-bake chapatti (R-BFC). The R-BFC samples were prepared from normal (control) and microwave treated wheat grains. The dough prepared from the control as well as treated whole wheat flour was sheeted into circular dough discs by pressing using a pneumatic chapatti pressing machine. The sheeted circular dough discs were packed in polyethylene pouches (75μ) with polypropylene lining (75μ) in between each dough disc and heat sealed. The samples were frozen to achieve core temperature of -18^{0}C and immediately stored at -18^{0}C for 6 months. The pouches after removal from the freezer were cut open and individual R-BFC were placed directly on the hot plate (230 ± 5^{0}C) and each side baked for 2 to 3 min to a golden brown colour. The texture profile analysis results showed that the extensibility of dough increased gradually in both the samples during frozen storage but with lesser increase in the case of treated samples. The increases in dough extensibility suggest weakening of the dough as the storage period advanced. The lesser increase in the extensibility of the treated R-BFC as compared with the control may be due to the inactivation of most of the lipolytic and proteolytic enzymes during microwave treatment of wheat grains. The maximum load for the extension of the dough also increased steadily in both samples as the storage period advanced and the increase was more in treated samples as compared with the control. This may possibly be due to the stiffening of the dough during frozen storage. The chapattis baked from both the R-BFC samples showed higher hardness, cohesiveness, chewiness and lower springiness values owing to the frozen storage effect on R-BFC. The data showed that springiness and chewiness of chapattis were correlated to the extensibility of the dough. The chapattis prepared from the microwave treated R-BFC samples were rated better in terms of colour, texture and overall acceptability scores as compared with the control up to 6 months of storage at -18^{0}C. Both the R-BFC samples remained microbiologically safe throughout the frozen storage period. Among the additives used to overcome the negative effect of frozen storage and repeated freeze-thaw cycles on the R-BFC, glycerol (1.0 per cent) was found to minimize the rheological changes, and improve the textural and sensory properties of R-BFC (Yadav et al, 2008c). The studies revealed that the procedure worked out for the storage of frozen R-BFC may popularize the frozen dough technology not only for the preparation of the staple traditional Indian flat bread (chapatti) but also the related products.

Among the processes discussed above to extend the shelf life of chapatti and related products, retort processing and frozen storage have been preferred as these processes make no or minimal use of preservatives. Retort processed and frozen stored chapattis, *parathas* and naan are being presently marketed successfully and are popular.

Chapter 8

Chapatti-related Products

Introduction

There are several culinary variants of chapatti; the important ones being *rumali roti*, *tandoori roti*, *missi roti*, *rotlee* of Gujarat, and *dopatris* of Rajasthan. The fat containing products which after rolling out, are shallow pan-fried using little fat include *paratha* or *parota*. The other variants that are deep fat fried include *poori*, *luchis* of Bengal, and *kachori* of Rajasthan. The fermented products include *bhatura*, *kulcha*, *khameeri roti*, *naan*, etc. These may replace chapatti in the dietary regimen of consumers who eat chapatti three times a day. However, *tandoori roti*, *poori*, and *paratha* more frequently replace chapatti. These flat breads differ based on the type of flour for the dough, the other ingredients the dough contains, filling or stuffing that goes in or the way they are prepared and served (Achaya, 1994). Except for *tandoori roti,* South Indian *parotta,* North Indian *paratha, poori,* and *naan*, the published information on majority of the products is scanty.

Tandoori Roti

Traditional Method of Preparation: *Tandoori roti*, unleavened, single layered flat bread is normally prepared using whole wheat flour or resultant *atta*. It is baked in a specially designed oval shaped oven called *tandoor*. The *tandoor* is made up of an oval earthen or a metallic pot placed vertically in-ground or over the ground, the walls of which are plastered with clay or built with refractory bricks. It is fired from the bottom side by burning wood or charcoal or by using natural gas. A piece of dough of suitable size (~50g) is rounded between the palms of hands and flattened by pressing and repeated slapping between the palms of hands to a circular dough sheet/disc of desired thickness (~4 mm) and diameter (~20 cm). The dough sheet is then supported on a cloth pad and pasted on the hot surface of the interior wall of the oven to bake. As soon as it attains creamish golden colour with light brown spots on the surface, the *tandoori roti* is taken out with the help of two iron skewers of suitable length and design. In Pakistan *tandoori roti* is a leavened product made from whole wheat flour, yeast and baking soda (Hashmi and Wooton, 1995).

Standardization of Method of Preparation: Tandoori roti represents a sizeable share of wheat based products consumption in the Indian subcontinent and has potential for increased consumption if produced and marketed in unit packs at commercial level. As a preamble to initiating standardization studies on various technological aspects of the preparation of *tandoori roti*, Saxena and Haridas Rao (1995) conducted a survey to assess the quality of flour, dough and *tandoori roti* from different sources in and around the city of Mysore, Karnataka, India. The study indicated that the consistency of *tandoori roti* dough was much softer than that of the chapatti dough and that the formulations invariably contained common salt to taste. The salt, apart from imparting taste, brings out the flavour of the *tandoori roti*, and improves the dough handling properties. It helps retention of moisture and retards staling. The *tandoori roti* samples collected from different sources had on an average 4.5 mm thickness and a diameter of 20 cm, and had wide variation in the texture.The *atta* used was either *chakki* whole wheat flour or roller mill resultant *atta*. It was generally found that the *tandoori rotis* prepared from whole wheat flour had better texture and flavour in comparison with those prepared from the mill *atta*. But the whole wheat flour *tandoori roti* is generally of duller appearance due to the high level of bran present in it. The commercial catering establishments prefer to use mill *atta* as it yields bright and attractive *tandoori rotis,* which the consumer relates to freshness. Moreover, the mill *atta* is economical to use as it is relatively cheaper than the whole wheat flour.

In order to standardize the method for the preparation of *tandoori roti* using whole wheat flour, Saxena and Haridas Rao (1996a) studied the relationship between ingredient based variables such as water and salt, and processing variables such as dough mixing time, baking temperature and time, which affect the quality of *tandoori roti*. The optimum conditions that promoted production of *tandoori roti* with maximum sensory score were:

- water content of 720 ml/ kg flour (corresponding to farinograph consistency of 800 BU at lever position 1:1),
- salt content of 11g/ kg flour,
- dough mixing time of 3.7 min in a Hobart mixer at 58 rpm,
- baking temperature of 425°C, and
- baking time of 37 seconds.

The maximum sensory scores for *tandoori roti* were based on the blister size, colour, appearance, handfeel, texture, mouthfeel, taste and aroma. In order to predict the product quality of *tandoori roti*, a model was developed using response surface methodology for determining the optimum processing conditions.

During baking of *tandoori roti*, the mode of heat transfer was found to be more important than just supplying the required quantity of heat for obtaining the desired characteristics in the baked product (Saxena et al, 1995). Conduction and radiation were found to be the main modes of heat transfer in the earthen *tandoor* and contributed to the characteristic crust, colour, texture and flavour of *tandoori roti*. Conduction contributed more to the product quality parameters when compared to other modes of heat transfer. However, the synergistic effect of different modes of heat transfer cannot be ignored.

Factors affecting the *Tandoori roti* Quality: The wheat variety and the method of grinding whole wheat flour may also influence the quality of *tandoori roti* in the same manner as for chapatti.

Effect of Wheat Variety: Saxena et al (1997) studied the rheological characteristics of dough, total protein and gluten proteins, and high molecular weight subunits composition of eight commercial wheat varieties (grown at one location) in relation to their *tandoori roti*-making quality. They observed wide variation in *tandoori roti* baking performance of these wheat flours. Four of these varieties viz. Cpan 3004, GW180, K 8804 and WL1562 were found suitable for the preparation of *tandoori roti*. These four varieties had higher farinograph water absorption capacity between 718 and 735 ml / kg when measured at an optimum *tandoori* dough consistency of 400 BU. The cohesiveness and extensibility of the dough from these varieties were also high compared to other varieties. These four varieties contained maximum amount of total protein, gluten as well as total glutenin. It was found that the varieties having protein content of about 10 per cent and total glutenin content of about 5 per cent, Glue1 protein subunit, and quality score of 6 produced good quality *tandoori roti*. The prediction equation derived by the authors for correlating total protein and total glutenin in flour with overall sensory score of the *roti* was found highly significant. The results suggested that the medium hard wheats are suitable for the preparation of *tandoori roti*.

Effect of Milling Method: Studies were also conducted to assess the effects of wheat milling methods (plate mill, hammer mill, pin mill or roller mill) used for the preparation of whole wheat flour (WWF) on the quality of the flour and of the *tandoori roti* prepared thereof (Saxena and Rao, 1997). The WWF quality characteristics were determined in terms of particle size distribution, composition, enzyme activities, amylograph properties and rheological properties of dough. The damaged starch content, diastatic activity and water absorption capacity were greatest in plate milled flour and lowest in roller milled flour. The dough made from the plate milled flour was more viscous than the dough made from the flours from other mills. The quality characteristics of *tandoori roti* were also studied in respect of moisture and total sugar contents, shear value, colour and sensory properties. The *tandoori roti* prepared from plate milled flour had soft texture and a sweetish taste, and was preferred over the other samples. The suitability of the flours milled in other mills for the preparation of *tandoori roti* decreased in the order: hammer mill > pin mill > roller mill.

Effect of Damaged Starch: The studies on the effect of damaged starch on the pasting characteristics and *tandoori roti* making quality of WWF revealed that the damaged starch content was significantly correlated with peak viscosity, and overall quality of *tandoori roti*. The overall quality of *roti* was mainly influenced by the sensory texture. The optimum damaged starch content of WWF for the preparation of good quality *tandoori roti* was found to be between 14 and 15 per cent (Saxena and Rao, 2000).

Effect of Carbohydrate Profile of Wheat: A study evaluated the carbohydrate profile of six wheat varieties differing in *tandoori roti*-making quality and found that the carbohydrate profile of the variety Cpan3004 was markedly different from those of other varieties. The variety Cpan3004 was found most suitable for making *tandoori roti* of good quality. The shear value of *roti* was inversely related to the sensory score. The study also revealed that arabinoxalans appear to play an important role in *tandoori roti*-making quality of wheat (Saxena et al, 2000).

Scope for Mechanization of *Tandoori Roti*-making

The earthen *tandoor* has a limited capacity as only 10-15 *tandoori rotis* can be baked at a time. Therefore, it is difficult to prepare this product on the large scale needed for the commercial preparation and marketing as done for other bakery products. To explore the possibilities of baking *tandoori roti* on commercial scale,Saxena and Haridas Rao (1996b) studied the effect of baking in different types of ovens such as an earthen *tendoor*, a gas *tandoor*, an electric oven and a heated plate, on the physicochemical and sensory quality characteristics of *tandoori roti*. The studies indicated that the extent of gelatinization of starch varied depending on the type of oven used and was found to be 74.6, 78.4, 90.7 and 88.4 per cent in *rotis* baked in an earthen *tandoor*, a gas *tandoor*, an electric oven and a heated plate, respectively. The pasting characteristics of starch, separated from the *tandoori roti* samples and measured using a Rapid Visco Analyzer, indicated the highest peak viscosity for *tandoori roti* baked in an earthen *tandoor* and the lowest for that baked in an electric oven. The concentration of high molecular weight proteins was lower in *roti* baked in an earthen *tandoor* and a gas *tandoor* indicating greater dissociation of high molecular weight proteins during baking in such ovens. The sensory texture, taste and flavour of *roti* baked in an earthen *tandoor* were found to be superior to those baked in other types of ovens. However, *tandoori rotis* baked in an electric oven were also found to be acceptable. The authors concluded that electrical oven after modifications in the design, may enable continuous production of good quality *tandoori rotis*.

North Indian *Paratha*

North Indian *paratha* or simply *paratha* or *parantha* is unleavened layered flat bread generally made from whole wheat flour (*atta*). *Paratha* is one of the most popular Indian unleavened flat breads, prepared either plain or stuffed. The plain *paratha* can be round, triangular or rectangular in shape. The stuffings / fillings may include spiced grated/mashed vegetables, cooked lentils/beans, *paneer* or any comminuted non-vegetarian product. *Parathas* are relished as a breakfast item in most of the Indian homes and are also served in restaurants, hotels and *dhaabas* on demand. Plain *paratha* is normally prepared from whole wheat flour with common salt to taste, shortening and water. The shortening generally used may be refined vegetable oil or ghee (clarified butter). After mixing, the dough is converted in to balls (*paeras)* of suitable size and each ball is sheeted by slapping between the palms of hands or on the board with rolling pin into a circular disc thicker than chapatti. The liquid shortening is applied on the surface of the dough disc and then folded several times and flattened again to a desired thickness. Shortening is applied on each fold to create discreet layers or laminations. Baking is done on preheated *tawa* (iron griddle) on both sides with intermittent application of the shortening. Baking may also be accomplished in preheated *tandoor* as done for *tandoori roti*. There are umpteen numbers of ways to create multiple layers to get *parathas* with soft and flaky texture. For example preparation of *lachha paratha* requires application of oil or ghee on the surface of the thinly rolled chapatti followed by lightly sprinkling fine wheat or rice flour, folding back and forth into several pleats, stretching the resulting strip and then coiling into a round thick disc. The disc is then gently rolled flat and baked on a preheated *tawa* with continued shallow frying by applying oil or ghee and frequently turning the upside down. The process of creating multiple layers or laminations may be mechanized using a laminated dough technique similar to the one used for flaky puff-pastry.*Lachha paratha* has an appetizing look with crumbly, melt-in-mouth texture.

Nutritive Value of *Paratha*

The nutritive value of plane *paratha* is slightly lower than chapatti and *poori*. *Paratha* is reported to differ significantly in phytic phosphorus content from other Indian flat breads resulting in lower absorption of zinc and other minerals from *paratha* diets (Grewal and Hira, 2003). The protein quality of *paratha* is slightly lower than that of chapatti as indicated by slightly lower values for true digestibility and net protein utilization of *paratha* compared to chapatti (Parimala and Sudha, 2015). On the other hand, stuffedparatha may have enhanced level of nutrients depending on the nature of the stuffing used. These authors have also reported that the vitamin A content of *paratha* may be enhanced by preparing *paratha* using deacidified and deodorized palm oil instead of other refined oils.The nutritive value of the commercially prepared frozen whole wheat flour North Indian *paratha* as per USDA Database (2016) is presented in **Table 1**.

Table 1: Nutritive value (per 100g) of commercially prepared frozen whole wheat flour North Indian *paratha*

Nutrient	Value, g	Nutrient	Value, mg	Nutrient	Value, mg
		Vitamins		**Minerals**	
Protein	6.36	Thiamin	0.11	Calcium	25
Fat	13.20	Riboflavin	0.076	Iron	1.61
Dietary fibre	9.6	Vitamin B_6	0.08	Magnesium	37
Sugars	4.15	Vitamin E	1.35	Potassium	138
Carbohydrates	45.36	Niacin	1.83	Phosphorus	120
		Vitamin K	3.4 µg	Sodium	452
				Zinc	0.82

Source: USDA Nutrient Database (2016).

Improving the Rheological Properties and Nutritive Value of North Indian *Paratha*

- **Supplementing with Multigrains:** Indrani et al (2011) studied the effect of multigrains on the rheological and quality characteristics of north Indian *paratha.* The multigrains blend (MGB) used consisted of flours of split chickpea (without husk), barley, soybean, and fenugreek seeds in the ratio of 4: 3: 2.5: 0.5, respectively. North Indian *paratha* (NIP) was prepared by supplementing whole wheat flour with 0, 10, 20, 30 and 40per cent MGB and a combination of additives (CA). The blends were mixed with salt (0.5 per cent), oil (5.0 per cent) and water (based on farinograph water absorption) in a Hobart mixer for 4 min at 61 rpm. The CA comprised dry gluten powder, sodium stearoyl-lactilate and hydroxypropylmethylcellulose at the rate of 2.0, 0.5 and 0.5 per cent, respectively. The dough was rested covered with wet cloth for 20 min. About 30 g of dough was rounded, sheeted to a thin circular shape using a wooden roller pin. About 2.5ml refined vegetable oil was applied all over the surface of the circular dough, folded into half circle, again applied about 2.5ml of the oil, folded into a quarter circle and then sheeted into a triangular shape of 3.0 mm thickness. The triangular dough disc was baked on a hot plate for 2 min at 180^{0}C while turning at every 30 sec to complete baking. The addition of increasing amount of MGB was found to increase water absorption, decrease the strength, elasticity, extensibility and amylograph characteristics of the whole wheat flour dough, and overall quality of the *paratha.* The decline in both breakdown and set back values of dough viscosity with the addition of MGB indicates less resistance of starch granules to thermal treatment and mechanical shearing, and less tendency to retrograde. The results presented in **Table 2** show that the blend with 30 g/100g MGB plus CA had increased farinograph water absorption, dough stability, extensograph resistance to extension, extensibility, and amylograph characteristics of whole wheat flour dough, and improved the overall sensory quality of north Indian *paratha.* The changes in physical and organoleptic properties of *paratha* with the addition of MGB and CA are shown in **Table 3**. The addition of multigrains increased the protein, fat, dietary fibre and mineral content of the North Indian *paratha* as well.

Table 2: Effect of multigrain blends (MGB) on the rheological characteristics of whole wheat flour

Parameter	**MGB(g/100g)**					
	0	10	20	30	40	30+CA
Farinograph						
Water absorption (g/100g)	74.6	74.8	75.0	75.2	75.5	77.0
Dough development time (min)	5.5	5.5	6.0	6.0	6.5	7.0
Dough stability (min)	4.8	4.5	4.0	3.5	2.5	6.5
Extensograph						
Resistance to extension, R(BU)	350	325	310	290	280	360
Extensibility, E (mm)	140	130	122	115	101	137
Amylograph						
Gelatinization temperature (^{0}C)	71.0	71.5	72.0	72.4	73.6	72.5
Peak viscosity (BU)	571	537	445	378	306	444
Breakdown (BU)	193	178	135	99	73	122
Setback (BU)	353	318	287	276	234	321

Source: Indrani et al (2011). CA: Combination of additives

Table 3: Effect of multigrain blends (MGB) on the quality of North Indian *paratha*

Parameter	**MGB (g/100g)**					
	0	**10**	**20**	**30**	**40**	**30+CA**
Physical						
Horizontal axis (mm)	115	108	106	104	100	116
Vertical axis (mm)	133	123	115	113	112	126
Thickness (mm)	3.5	4.0	4.2	4.4	4.6	4.0
Shear force (g)	1265	1300	1350	1400	1570	1310
Sensory						
Appearance (10)	9	9	8	8	7	8.5
Pliability (10)	9	9	8	7	6	8.5
Tearing strength (10)	8	7	7	6	5	8.5
Layer separation (20)	18	17	16	15	13	17
Eating quality (10)	9	9	8	8	7	8.5
Overall quality (60)	53	51	47	44	38	51

Source: Indrani et al (2011). CA: Combination of additives

- **Supplementing with Barley Flour, Banana Flour and Soy Protein Isolate:** For commercial production of nutritious North Indian *paratha* (NIP), Bhargava et al (2012) supplemented whole wheat flour with blends of barley flour, banana four and soya protein isolate. The blend was found to influence the rheological and nutritional characteristics of NIP. There was an increase in the farinograph water absorption, amylograph peak viscosity and extensograph resistance to extension, and a decrease in extensibility. The overall quality score of NIP decreased from 54 to 25 (maximum score 60) with increased addition of the supplements up to 60 per cent. The addition of combination of additives (dry gluten powder, sodium stearoyl-2-lactylate and hydroxypropylmethylcellulose) significantly improved the overall quality of NIP with 40 per cent supplements. The NIP with supplements (40 per cent) and additives showed higher total protein, digestible protein and percent availability, total fiber, soluble and insoluble fibers, resistant starch but lower starch hydrolysis index as compared to the control NIP. The use of these supplements may, therefore, be exploited in the commercial production of nutritious North Indian *paratha*.

Preservation, Packaging and Storage of *Paratha*

Kameswara Rao et al (1969) preserved the plane paratha satisfactorily for 10 months by incorporating sorbic acid and common salt in the dough. Storage studies under diverse tropical conditions showed that initial packing in MST Cellophane and repacking in heat-sealed paper-foil-polythene pouches kept the *parathas* soft and supple during the storage period. However, there was slight increase in fat acidity. The reducing sugars increased at the expense of non-reducing sugars. Overall, the preserved *parathas* were found acceptable during the entire storage period.

Mathur, et al (1973) could preserve *parathas* in acceptable condition for more than a year by processing in cans at 116^0Cfor 100 minutes. However, the quality of thermally preserved *parathas* was found far inferior to the one preserved by sorbic acid due to resultant excessive caramelization, and hard and leathery texture.

Kannur et al (1972), however, preserved the stuffed (vegetarian or non-vegetarian curried stuffings/fillings) *parathas* by heat processing in flexible pouches. The stuffed *parathas* singly or in pair were first packed in MST Cellophane pouches and then in paper-foil-polythene laminate pouches, and finally heat sealed. The sealed pouches were then processed in a specially designed autoclave under steam pressure (15 psi) for desired period. After processing steam was cut off and air was let in to achieve a pressure of about 20 psi. Water was then sprayed on the pouches to effect cooling. After draining out the water the air pressure was maintained for another two minutes and then released slowly to prevent bursting of pouches. The in-pack sterilized *parathas* with varieties of stuffings / fillings were found acceptable up to 6 months storage.

Khan et al (2014) studied the thermal processing and shelf-life of potato stuffed *parathas*. The potato stuffed *parathas* were prepared from whole wheat flour dough using cooked and mashed potatoes seasoned with required quantity of spices and shallow fried in hydrogenated fat at 120 to 130^{0}C. The response surface technology was utilized for the optimization of ingredients like dough and potato stuffing. The baked stuffed *parathas* were packed in indigenously developed retortable pouches and processed in a steam-air retort. The total processing time was 25 min at 121±1^{0}C to achieve a F_0 value of 3.5. The quality of the product was evaluated chemically, sensorially and microbiologically to assess the shelf-life. Even though, there was an increase in the peroxide value, free fatty acid value and thiobarbituric acid value, still the product remained stable and acceptable for 12 months as the product was microbiologically stable during the entire period of storage. However, after 12 months of storage, the sensory scores decreased significantly from 8.6 to 7.1.

Bhoir et al (2015) developed shelf-stable ready-to-eat (RTE) m*ethi paratha* (flavoured Indian unleavened flat bread) using combination of hurdles including vacuum packaging and radiation processing under frozen conditions. The m*ethi paratha* was prepared using dough containing whole wheat flour, dried *methi* (fenugreek leaves), vegetable oil and spices (red chilli powder, cumin seeds, carom seeds, ginger-garlic paste and turmeric powder). The *parathas* were vacuum packed in multi-layered pouches and irradiated at 25 kGy in frozen condition. The samples were evaluated for microbiological quality, lipid peroxidation and sensory attributes during storage under ambient conditions (25 to 30^{0}C). The samples were found free of any viable microorganism throughout the storage period of 180 days. The thiobarbituric acid reactive substances value which indicates lipid peroxidation of samples did not show any significant increase with time. The m*ethi paratha* remained sensorially acceptable during the entire period of storage.

SOUTH INDIAN *PAROTTA*

Introduction

The populace of South India though predominantly rice eating, there has been substantial increase in the consumption of wheat during the last few decades. The traditional wheat-based foods that have gained popularity include *chapattis/phulka*, naan, *bhatura* and South Indian *parotta* (Indrani, 1998). These products utilize more than 50 per cent of the wheat flour (*maida*) produced by the roller flour mills in South India. The South Indian *parotta* is also known as Kerala *parotta*, Ceylone *parotta* or simply *parotta*. It is circular shaped, multilayered, creamish white coloured, unleavened flat bread made essentially from white wheat flour (*maida*) unlike North Indian *paratha* which is prepared from whole wheat flour (*atta*). It has been one of the staple food items in the southern states of India comprising, Andhra Pradesh, Karnataka, Kerala and Tamil Nadu.

Traditional Method of preparation

Traditionally, the preparation of *parotta* involves the use of white wheat flour (*maida*), common salt to taste, water, and refined vegetable oil to facilitate thin sheeting of the dough (Indrani and Venkateswara Rao, 2001). However, sugar and egg are also used as optional ingredients perhaps to improve the taste and texture of the final product. The dough prepared from the desired ingredients is sheeted using a rolling pin into a thin film by applying oil intermittently, folded horizontally into multiple layers and then coiled. The coiled dough sheet is finally sheeted into a circular disc of desired thickness and diameter, and baked on a hot griddle (*tawa*) with frequent turning upside down and applying edible oil at the same time.

Desirable *Parotta* Quality Characteristics

A *parotta* with creamish white colour having light brown spots scattered all over the surface, a circular shape with distinct layers, a soft and pliable handfeel, a soft and slightly chewy texture, optimum oiliness, easy breakdown in the mouth, and typical pleasant aroma and taste is considered to be highly desirable (Indrani et al, 2000). The survey of the quality characteristics of *parotta* collected from different commercial sources showed variation in the overall quality score from 40 to 75 out of maximum score of 100 indicating inconsistency in the quality of *parotta* as judged by the taste panel members who were familiar with the product (Indrani and Venkateswara Rao, 2004). This prompted the authors to develop a test baking method to prepare *parotta* of optimum quality.

Test Baking Method

The development of a test baking method for the preparation of *parotta* necessitated the need for standardization of ingredients used and the process conditions for the preparation of *parotta* of desirable quality characteristics as shown in **Table 4.**

Table 4: Quality characteristics of *parotta*

Characteristic	Desirable	Undesirable
Colour	Creamish white	Brownish white, dark brown
Nature of spots	Uniform, light brown	Non-uniform, dark brown, chard
Shape	Circular	Non-circular
Layers	Distinct layers	Less distinct, completely fused layers
Handfeel	Soft and pliable	Hard
Texture	Soft and slightly chewy	Highly chewy, chewy, tough
Oiliness	Optimum	Oily, dry
Mouthfeel	Easy breakdown, no residue	Sticky, doughy
Taste and aroma	Typical, pleasant	Foreign, bland, off flavour

Source: Indrani and Venkateswara Rao (2004).

The formulation derived from the experiments on optimization of ingredients by using response surface methodology (Indrani and Venkateswara Rao, 2001) contained wheat flour (*maida*) 100g, water 53.6 ml common salt 1.0 g, sugar 0.8 g, egg 9.6 g, and oil for sheeting 16.6 g.The farinograph dough consistency of 500 BU was found optimum to prepare *parotta* with a maximum overall quality score of 75 out of 100. The studies on the variability in dough mixing and rest periods, dough sheeting thickness, *parotta* baking time and temperature assisted in the development of optimum *parotta* processing conditions. The optimum processing conditions used for the preparation of *parotta* were (Indrani and Venkateswara Rao, 2004):

Optimum processing condition	Value
Dough mixing time	4.0 min
Rest period	30 min
Sheeting thickness	5.0 mm
Baking temperature	230°C
Baking time	2.0 min

The dough was prepared by mixing all the ingredients for 4.0 min in the Hobart N:50 mixer at 61 rpm using a J-hook blade. The dough covered with a wet cloth was rested for 30 min in a cabinet at 30^0C and 75 per cent RH. The dough was divided into pieces of 75g each, rounded and again rested in the cabinet for 10 min. The rounded pieces were then sheeted using a rolling pin. The oil was applied during rolling to facilitate formation of a thin dough film. The thin dough film was folded from the edges in the horizontal direction to form multiple folds and then coiled. The coiled dough sheet was rested for 10 min and sheeted again using a guide frame to get the *parotta* dough sheet with a thickness of 5.0 mm and diameter of 150 mm. The *parotta* dough sheet was baked on a thermostatically controlled hot plate at temperature of 230^0C for 2 min. During baking the *parotta* was turned upside down every 15 sec while applying a total of 5.0 g oil. The *parotta* was then cooled to room temperature before quality check. The overall quality score of *parotta* prepared from the flours varying in protein quantity and quality, dough mixing time and stability was highest at 4.0 min dough mixing time.

Prabhasankar et al (2003) used scanning electron microscopy and electrophoresis techniques to study the changes that occur in starch and protein molecules during processing of *parotta* starting from wheat flour to baked *parotta*. At the processing stage of *parotta* dough into a very thin sheet, the surface of dough appeared as a fine matrix of proteins and soluble solids with dispersed starch granules while the surface of baked *parotta* showed a still finer distribution of all the components. Electrophoretic patterns of wheat flour, *parotta* dough and baked *parotta* revealed that less intense bands were observed in glutenin HMW regions of baked *parotta*. These interactions between starch and protein matrix perhaps explain the typical textural characteristics of baked *parotta*.

Factors Influencing the Quality of parotta

Effect of Wheat Flour Characteristics: Indrani and Venkateswara Rao (2000) analyzed six wheat flours differing in quality and identified chemical and rheological characteristics that influence the quality of *parotta*. Among the chemical characteristics, quantity and quality of protein was found to be the best index in predicting the quality of *parotta*. The dry gluten content was greatly correlated with colour while the nature of spots on the surface of *parotta* was correlated with the damaged starch content. The flour protein content showed highest correlation coefficient with sensory scores, indicating that the protein content was the prime factor influencing the quality of *parotta*. Among the rheological properties such as farinograph water absorption, extensograph ratio figure and area, and Instron apparent biaxial extensional viscosity, hardness, and cohesiveness were found to be highly correlated with the overall quality score of *parotta*. The sensory texture of *parotta* was found to be highly correlated to the instrumental shear and compression forces showing that shear and compression forces could be considered as the best indices of typical sensory texture of *parotta*.

Indrani and Sai Manohar (2007) reported the usefulness of the alveograph in predicting the quality of wheat flours for the preparation of *parotta*. The alveograph index of swelling (G) and deformation energy of dough (W) were found highly correlated with the spread ratio, texture, layers and overall quality score of *parotta* indicating that alveograph G and W values could be considered as the indicators of the overall quality of *parotta*.

Effect of Flour Particle Size: The wheat flour fractionated into different particle sizes was found to exhibit variation in chemical composition, rheological properties and sensory quality of *parotta* prepared from it (Sakhare et al, 2014). The finer fraction (<75 and 75 to 118 μm) had lower ash, higher dry gluten, damaged starch and SDS sedimentation value than the coarse fraction (118 to 150 and > 150μm). The finer fractions exhibited stronger quality gluten and dough exhibited better balance of elasticity and extensibility compared to other coarser fractions with higher dough development time and dough stability. The *parottas* prepared from the finer fraction showed significantly higher sensory scores for colour, texture, layers, mouthfeel and overall quality score than the coarser fractions. These studies indicated that the wheat flour from the finer fractions yields higher quality *parotta* owing to the presence of higher damaged starch content, quantity and quality of protein in these fractions than coarser fractions.

Effect of Wheat- Flour Mill Streams: With the objective to develop specialty flour for *parotta*, Indrani et al (2003) studied the chemical, rheological and *parotta*-making properties of wheat-milling flour streams. The *parottas* made from the first five break passages had a decreased spread ratio, dull brown colour, and fused layers and lower overall quality score. The initial reduction streams (C1 to C5), on the other hand, produced good quality *parottas* in terms of appearance, spread ratio, layers and texture. The speciality flour made by combining C1 to C5 streams had 0.45 per cent ash, 9.63 per cent dry gluten, 0.7 per cent free and 1.8 per cent bound lipids, 45 ml sodium dodecyl sulphate sedimentation value, and produced parottas having creamish white colour, excellent pliability, thin, transparent and distinct layers, soft texture and moderate chewiness with easy dissolution in the mouth without leaving any residue.

The overall quality score of parotta was highest (93.5) for the specialty flour when compared to break (43 to 77) and reduction (56 to 92) streams.

Influence of different Additives: The rheological characteristics of wheat flour dough and quality of *parotta* as influenced by the use of different additives has been investigated by several authors (Indrani and Venkateswara Rao, 2003; 2006 and 2007; Prabhasankar et al, 2004; Smitha et al, 2008; *Tejaswi kumar et al, 2013*).

The effect of different levels of ***parotta* ingredients** such as salt, sugar, egg and refined groundnut oil on various rheological properties of wheat flour dough was studied using farinograph, mixograph, extensograph and Instron universal testing machine (Indrani and Venkateswara Rao, 2007). Addition of salt or egg was found to increase the strength and elasticity of the dough, whereas addition of oil decreased the strength and viscosity of the dough.

Indrani and Venkateswara Rao (2003) also tested the effect of **different surfactants** including glycerol monostearate (GMS), diacetyl tartaric acid esters of monoglycerides (DATEM), polyoxyethylenesorbitanmonostearate (PS-60), lecithin and sodium stearoyl-2-lactylate (SSL) on the rheological characteristics of dough and quality of parotta. Use of the surfactants increased stability as measured by the farinograph, valorimeter value, extensograph ratio figure and extensograph area, indicating an increase in the strength of the dough. Surfactants reduced the mixograph peak height, mixograph area, apparent biaxial extensional viscosity, compressive stress, force decay parameter, hardness and adhesiveness, and increased the cohesiveness of dough. The values for peak viscosity increased with DATEM, PS-60 and SSL and decreased with GMS and lecithin. SSL and PS-60 were found to bring about the greatest improvement in the quality of parotta, followed in decreasing order by DATEM, lecithin and GMS.

Enzymes are commonly used in the baking industries to improve dough properties and the quality of end products. Prabhasankar et al (2004) studied the role of **different enzymes** such as fungal α-amylase, glucose oxidase, proteinase and xylanase in the baking process of South Indian parotta. The rheological characteristics and microstructure of parotta dough in relation to the overall quality characteristics of baked parotta was investigated. The effect of different enzymes on the rheological properties such as farinograph stability, extensograph resistance to extension, and the extensibility was found variable. The microstructure of parotta dough and baked *parotta* with different enzymes revealed that the use of proteinase improved the continuous gluten film formation compared with the control and the other three enzymes. This could be due to the proteinase enzyme breaking larger protein fibrils into smaller fibrils and thus enhancing the continuous gluten film formation. Interaction of protein films and starch granules was also affected differently with the use of different enzymes. The study reported that the overall quality score of parotta decreased to 68 and 64 with the use of fungal α-amylase and glucose oxidase, respectively as against the control parotta score of 74. However, when proteinase and xylanase were used, the overall quality score increased to 92 and 82, respectively. Thus, the addition of proteinase or xylanase to parotta dough was found to be more effective in improving overall quality of South Indian *parotta*.

Indrani and Venkateswara Rao (2006) investigated the effects of **oxidizing agents** (potassium bromate, ascorbic acid and potassium iodate), reducing agents (potassium metabisulphite and cysteine hydrochloride), enzymes (fungal á-amylase and protease), and dry gluten powder on the rheological characteristics of dough and quality of parotta. These additives were found to modify the rheological characteristics of wheat flour dough to varying extent. The rheological modifications brought about by separate incorporation of 100 ppm potassium metabisulphite, 50 ppm of cysteine hydrochloride or 10 ppm protease resulted in yielding parotta of highly desirable quality. These observations will be useful in producing specialty wheat flour for parotta of desired quality.

The effect of **hydrocolloids** on the rheological, microstructural and quality characteristics of parotta as studied by Smitha et al (2008) showed increase in water absorption capacity of wheat flour dough with individual addition of Arabic (A), guar (GR), xanthan (XN), carrageenan (CG) and hydroxypropylmethyl cellulose (HPMC). The strength and elasticity of the dough increased with XN. The elasticity of dough was moderately increased by AR, GR, CG and HPMC. The dough was more extensible with AR and HPMC. The peak viscosity was highest with GR whereas setback was lower with XN. The microstructural characteristics of the dough with hydrocolloids showed that the starch granules were coated with gum. The coating was more evident in the case of dough with GR and HPMC. Maximum improvement in the overall quality of parotta was brought about by GR. During storage, parottas with xanthan gum not only remained softer than the control but also were chewy.

Tejaswi kumar et al (2013) investigated the effect of **different fats** separately at 7.5 per cent level on the rheological, fatty acid profile and quality characteristics of parotta. Addition of fats and oils decreased farinograph water absorption, amylograph peak viscosity, alveograph resistance of dough to deformation and increased average abscissa at rupture. In the micrographs of *parotta* dough with fats, the protein matrix appeared thick and intact, whereas in the case of *parotta* dough with oils the protein matrix appeared slightly less intact. The spread ratio and overall quality of *parotta* with the oils were higher than with the fats. The highest overall quality score was observed for *parotta* with olive oil, followed in decreasing order by sunflower seed oil, soybean oil, coconut oil, palm oil, hydrogenated fat and bakery fat. Determination of fatty acid profile showed that the *parottas* with fats contained a high amount of trans-fatty acids, while *parottas* with oils had no trans-fatty acids. The *parottas* prepared with oils remained softer than the *parottas* with fats during storage up to 48 h.

Production of Specialty Parotta

With the objective to produce specialty parotta, several supplements were used to increase the nutritive value or provide popular flavours and other functional properties to parotta (Indrani et al, 2007 and 2010; Kaushika Maya et al, 2015).

Indrani et al (2010) showed that replacing wheat flour with increasing levels of fenugreek seed powder (FSP) increased farinograph water absorption, amylograph peak viscosity, spread ratio of parotta but decreased extensograph extensibility and shear force. However, the values for farinograph stability, extensograph resistance to extension and area increased only up to 5 per cent FSP level. This level of supplementation had no adverse effect on the overall quality characteristics of parotta, and imparted likable perceptible fenugreek flavour.

Kaushika Maya et al (2015) studied the effect of increasing levels of barley flour on the rheological characteristics of dough and quality characteristics of south Indian *parotta*. The use of increasing amount of barley flour from 0 to 40 per cent increased farinograph water absorption and decreased stability, increased extensograph resistance to extension and decreased extensibility, and increased amylograph peak viscosity and decreased set back values. Sensory analysis data showed that use of barley flour above 30% brought about adverse effect on the quality of parotta. However, addition of combination of improvers, e.g. 2 per cent dry gluten powder and 0.001 per cent enzyme xylanase decreased elasticity, increased extensibility of the dough and quality of parotta with 30 per cent barley flour. The supplemented parottas remained softer than the control parotta during storage for 48 h.

Indrani et al (2007) studied the impact of addition of whey protein concentrate (WPC) up to 15 per cent level on the quality of parotta. Supplementation with WPC at 5 per cent level only, yielded parotta with soft texture, thin and transparent layers like that of control parotta. Beyond 5 per cent level of WPC, the quality characteristics of parotta were adversely affected.

Storage life of *Parotta*

Commercial production and marketing of parotta in unit packs will require that the parottashould have adequate shelf life. For extending the shelf life of any product, the knowledge regarding the changes occurring in the product during storage is necessary.Indrani et al (2000) studied changes in the physicochemical and organoleptic characteristics of *parotta* during storage. There was decrease in alkaline water retention capacity (AWRC), in total water-solubles (TWS), in soluble starch (SS), in solubilised amylose, and in solubilised amylopectin during storage of *parotta* from 0.25 to 48 h.The pasting characteristics of stored *parotta* measured using a Rapid Visco Analyzer also showed a decrease in maximum viscosity at 95^0C The overall quality score of *parotta* based on colour, nature of spots, shape, oiliness, handfeel, texture, layers, mouthfeel, taste and aroma decreased from 75 to 43 while the shear and compression force of parotta increased considerably with storage time. The overall quality score showed highly significant correlation with AWRC, SS, amylopectin and viscosity at 95^0C after 15 min. The differential scanning calorimeter (DSC) characteristics of stored *parotta* samples indicated an increase in the enthalpy of the endotherms with storage time, suggesting different degrees of starch retrogradation.

Ray and Dasappa (2017) used a combination of ingredients namely raisin paste, citric acid and vinegar collectively designated as RCV to study the effect on the rheological, physico-sensory, microbiological and storage characteristics of south Indian *parotta.* Addition of RCV decreased farinograph water absorption, dough stability, extensograph resistance to extension, extensibility; increased amylograph pasting temperature, peak viscosity, cold paste viscosity, set back, and decreased break down values. There was decrease in pH and $\mathbf{a}_w$ of baked *parotta* with increased overall quality score of *parotta.*The sensory evaluation of *parotta* during storage at ambient temperature (27 $\pm 2^0$C) showed that the *parottas* were acceptable up to the 9th day, suggesting the possibility of producing ready to eat, microbiologically stable *parottas* with a shelf life of 9 days.

The shelf life of thus preserved parotta may further be extended by freezing and storage at -20^{0}C or by par baking followed by frozen storage; alternatively freezing the *parotta* dough, then thawing and baking when required as done for chapatti (Gujral et al, 2008). Indrani et al (2002) studied the effect of storage of frozen *parotta* dough (FPD) and ready-to-bake frozen *parotta* dough-disc (RB-FPD) at -20^0 C for three months on the rheological and baking characteristics, respectively. The optimum thawing conditions for FPD and RB-FPD were found to be 1 min when thawed in the microwave oven or 16 h when thawed in the refrigerator. The rheological characteristics of FPD showed a decline in farinograph peak consistency and an increase in extensograph extensibility due to storage. The major protein fractions of FPD as shown by the electrophoretic patterns remained unchanged up to a period of three months. However, when compared to FPD, the protein bands in the high-molecular-weight region of glutenin were less intense for baked *parotta* from RB-FPD. The overall sensory rating of *parotta*s prepared from RB-FPD samples was good even after three months of frozen storage.

NAAN

Introduction and Ingredients for naan making

Naan is a traditional leavened flat bread prepared from refined wheat flour (*maida*) and is baked in a *tandoor* like *tandoori roti.* It is the much sought after flat bread in the Indian subcontinent and is popular in Central Asia and Southeast Asia. In India, naan is the favorite cuisine in *dhaabas* (roadside eateries), restaurants and hotels but is occasionally prepared only in some households. The ingredients that are commonly used to prepare naan include *maida*, curd (yoghurt), refined oil /fat, milk, egg, common salt, sugar and baking soda (Rahim et al, 1993). Milk, egg and sugar are optional. In India, naan is made from refined wheat flour of about 72 per cent extraction, whereas in Pakistan it is made from brown flours of extraction rates in the range of 88 per cent (Hashmi and Wootton (1995). The scientific reports related to naan are rather scanty.

Standardization of Method for Naan Making

Rahim et al (1993) after surveying the *naan* serving establishments for ingredients and processing conditions used for *naan* making, arrived at the optimum levels of ingredients and desired processing conditions to standardize a laboratory method for preparing dough of desired consistency and naan of good overall quality. The ingredients comprising refined wheat flour (*maida*) 100g, curd (pH 4.8) 12g, salt 1g, fat 4g, milk 6g, egg 3g, sugar 2g, baking soda 0.5g and quantity of water equivalent to naan dough water absorption (Research Water Absorption Meter extrusion time range: 65-70 sec) were mixed for 3 min in a Hobart mixer. The resultant dough was allowed to ferment for a period of 4 h. About 80 g of the fermented dough was sheeted to a desired thickness of 2.5 mm using a rolling pin and a specially designed platform frame. The naan dough sheet was baked in a gas *tandoor* set at maximum flame (350 to 365^{0}C) for optimum period of 1.5 min (One side of the sheet baked for 1 min on the inner side of the covering lid and the other side for 0.5 min on the wire mesh of the *tandoor* directly). The resultant naan had pleasing appearance, soft to touch, spongy to handfeel, somewhat firm to tear, a little chewy to bite without sticking to teeth, and possessed a characteristic fermented flavour and good overall acceptability. The typical characteristics of wheat flour used for naan making are listed in **Table 5.**

Table 5: Typical characteristics of wheat flour for naan making

Characteristic		Wheat flour	
		Commercial	Laboratory
Particle size (as throughs of 25 p sieve of 62μ)		18.6-22.7	21.6-31.7
Moisture	%	11.8-13.1	12.3
Wet gluten	%	21.7-23.9	29.6
Damaged starch	%	11.4-13.7	9.4-12.6
Total ash	%	0.59 - 0.80	0.48
Sedimentation value	ml	17.7-28.2	28.2
Pelshenke value	min	85-106	124
Flour hydration capacity	%	206-225	236
Farinograph water absorption	%	58.1-62.1	62.3
Hand mixed dough (NWA*)	%	61.0-65.1	64.5
Hobart mixer dough (NWA*)	%	59.8-64.1	63.2
Extrusion time	sec		
Hand mixed dough		40-55	54
Hobart mixer dough		65-76	76

Source: Rahim et al (1993). * Naan-dough water absorption

Nutritive Value of Naan

Nutritionally, naan is superior to chapatti and *roti* as it is prepared from fermented dough containing curd (yoghurt), milk and egg. The nutritive value of the commercially prepared and refrigerated naan as per USDA Database (2016) is presented in **Table 6.**

Table 6: Nutritive value (per 100g) of commercially prepared, refrigerated *naan*

Nutrient	Value, g	Nutrient	Value, mg	Nutrient	Value, mg
Carbohydrates	50.43	**Vitamins**		**Minerals**	
Sugars	3.55	Thiamin	0.75	Calcium	84.0
Dietary fibre	2.2	Riboflavin	0.51	Iron	3.25
Fat	5.65	Vitamin B_6	0.095	Magnesium	27.0
Protein	9.62	Vitamin E	0.79	Potassium	100.0
				Phosphorus	125.0
				Sodium	405.0
				Zinc	0.81

Source: USDA Nutrient Database (2016).

Dogra et al (2001), however, further improved the nutritional quality of naan by supplementing refined wheat flour with soybean flour obtained from differently treated soybeans. Supplementation with treated soybean flours up to 15 per cent level was found acceptable for making naan. Using response surface methodology, Farooq et al (2013) prepared naan of acceptable texture and sensory appeal from wheat flour supplemented with optimum level of gram (chickpea) flour (13.06 per cent) and 0.32 per cent glyceryl monostearate as emulsifier. These supplementation studies directly address the solution for protein malnutrition without deteriorating the appeal and acceptability of the product, naan.

Reports and Patents on naan

Honeytop Foods, UK (2000) has developed an advanced naan bread production facility to produce high quality; flame baked using similar method as practiced in restaurants and hotels. The processing stages and equipment include mixing of ingredients, moulding, proving at 30 to 40^{0}C for 1 h, rolling and sheeting, proving again for 30 to 40 min, shaping (oblong) by hand, flame cooking in a clay lined long travelling oven for less than 1 min, spiral cooling, packaging in modified or controlled atmosphere packaging. The product ranges include plane, *peshawri*, garlic, keema, *pakora*, coriander and onion *bhaji* flavoured naan, and a range of filled naans.

Ajmera et al (2015) patented a system and method for preparing naan bread. A conveyor oven was used to bake flattened and pre-shaped naan bread loaves. The conveyor oven uses high temperature gas heating to bake the naan imparting traditional characteristics such as the slightly smoky or scorched appearance.

POORI

Poori spelled as *poory* or *puri* as well, is an ancient traditional unleavened, deep-fat fried, puffed product of South Asia particularly India, Pakistan, Bangladesh and Nepal. It is usually prepared from whole wheat flour; all-purpose flour may also be used. It is a versatile product not only consumed as a complete meal but also relished as breakfast or as a snack item.

Traditional Method of Preparation of *Poori*

Traditionally, light and delicious *pooris* are prepared using coarsely ground whole wheat flour (*atta*). The dough is kneaded in the similar manner as for chapatti but little less water is used to make relatively firm dough. It is shortened with a little ghee (clarified butter) or refined oil and common salt to bring out the flavour. The process for the preparation of dough involves sifting *atta* and salt together in a mixing vessel. Melted ghee or oil and some warm water are added into the well made in the *atta* and mixed. Thorough mixing is accomplished by rubbing the flour mix between the palms and fingers. The remaining amount of water is added gradually and kneading continued till the dough is formed. The dough should be pliable but moderately stiff. After applying oil or ghee on the hands, the dough is kneaded again for about 5 min until firm and smooth. The dough is then kept covered to rest for 30 to 60 min after applying little oil on the surface. After resting, the dough is kneaded for a few seconds and set aside covered with a damp cloth to prevent drying. Small portion of the dough is taken and made into ball. The ball is rolled into a disc 1.5 to 2.5 mm thick and 10 to 12 cm diameter using a rolling pin. While sheeting, oil or ghee is applied on the surface of the board and the rolling pin to prevent sticking. Frying of *poori* is accomplished in preheated (~185^{0}C) ghee or refined edible oil. The *poori* dough disc is carefully slipped into the hot oil to avoid folding. It will first sink to the bottom of the pan and then quickly rise to the surface. As it begins to puff, press it very gently with the back of the perforated flat ladle until it puffs off completely. Hard pressing should be avoided to control any cracking of the surface that may otherwise allow the oil or ghee to go inside the *poori*. When the bottom of the puffed *poori* gets slightly browned, it is carefully flipped over upside down using the perforated ladle. It may take 45 to 50 sec to complete the frying to golden brown colour on both sides. The fried puffed *poori*is taken out with the help of the perforated ladle. The process of ball making, sheeting and frying is continued for the rest of the dough.

Standardization of Process of Making *Poori*

Shurpalekar and Shukla (1992) standardized the optimum conditions for the preparation of *pooris* from whole wheat flour (*atta*) and all-purpose flour (*maida*). The preparation of dough of desired consistency, and *poori* of optimum quality required:

- water absorption of 67±2 per cent for *atta* and 52 to 54 per cent for *maida*,
- rolling of dough sheet to 2 mm thickness and 12.5 cm diameter yielding *pooris* with full puffing,

- frying temperature of 180±2^{0}C, and
- frying time of 25 to 30 sec.

As the quantity of water used for the preparation of dough increased from 65 to 71 per cent, the oil pickup by *pooris* also increased from 29.1 to 34.4 per cent, while the moisture content of *pooris* decreased from 25.6 to 19.7 per cent. This perhaps contributes to crisp texture and better eating quality of the *pooris*. The texture and eating quality of *maida poori* was inferior to that made from *atta*, but improved on addition of 5 to 10 per cent bran; however it's puffing was affected adversely and oil uptake was more. It was also observed that water required for dough making as well as the oil uptake in *atta* poori were comparatively higher than those made from *maida* (**Table 7**).This may be attributed to the presence of most of the wheat bran retained in *atta*, unlike in the case of roller milled *maida*.

Table 7: Effect of water used in the dough and sheet thickness on the quality of *poori* based on *atta* and *maida*.

Water absorption (per cent)	Thickness of doughsheet (mm)	Height on puffing (cm)	Moisture in *poori* (per cent)	Oil pick up in *poori* (per cent)
–	–	- *Atta*-	–	–
65	2	5.2	25.40	29.72
67	2	5.5	22.64	29.75
69	2	5.5	22.50	32.47
71	2	5.6	20.51	34.37
65	1.5	4.0	25.57	29.11
67	1.5	4.0	20.56	31.17
69	1.5	4.2	19.67	32.61
71	1.5(could not be sheeted)	–	–	–
–	–	-*Maida*-	–	–
48	2	5.3	20.6	26.72
52	2	5.3	19.7	27.01
56	2	5.4	19.2	28.72
48	1.5	4.3	21.2	25.50
52	1.5	4.4	20.2	26.41
56	1.5	4.7	19.1	27.01

Source: Shurpalekar and Shukla (1992).

Vatsala et al (2001) used response surface methodology to optimize the level of ingredients, and process conditions for the preparation of whole wheat flour *pooris* of widely acceptable quality. The overall sensory scores based on colour, appearance, handfeel, texture, mouthfeel, taste and aroma; the shear value; and oil uptake were used to assess the product quality in the preparation of *pooris*. The optimum conditions which determined the maximum sensory score of 6 to 7 out of 10, minimum shear value of 3 to 3.5 N, and minimum oil uptake of 18 to 18.8 per cent were:

- flour extraction, 88.5 per cent,
- water absorption, 65.1 ml per 100 g flour with corresponding farinograph consistency of 450 BU at lever position 1:3;
- dough resting period, 28.8 min;
- *poori* frying temperature, 200°C, and
- frying time, 37.5 sec.

Pooris of good overall acceptability could also be prepared from composite flours based on wheat *atta* blended with maize / *jowar* (sorghum) flours at 15 per cent level or with Bengal gram flour at 20 per cent level. The Bengal gram flour supplemented *pooris* possessed better texture than that of *pooris* containing maize or *jowar* flour (Shurpalekar and Shukla, 1992). The utilization of such composite flours may be considerably beneficial for *poori* consuming populations in parts of the world where maize and *jowar* are preferentially grown and wheat products are imported.

Mechanization of *Poori* Making

Traditionally, *poori* dough is sheeted manually using a rolling pin, which is cumbersome and unhygienic. Recently, Yadav and Rajan (2012) used a manual chapatti pressing machine to make a circular *poori* dough sheet of desired thickness.

In order to meet the rising demand for ready-to-eat *poori,* mechanization of *poori* preparation has been explored (Venkateshmurthy et al, 2015) using an improved automatic device for pneumatic extrusion of dough into sheet and shaping, cutting of extruded sheet for preparation of *poori* and other similar Indian traditional foods (Venkateshmurthy et al, 2008). Using this device to automatically extrude *poori* dough sheet, processing conditions were optimized to get *poori* of maximum overall sensory quality with minimum shear force and minimum oil uptake. The moisture content of the *poori* dough is reported to play a very important role in this context. Also, the pneumatic pressure, rate of extrusion, frying time and puffed height of the *poori* are the other parameters that depend on the moisture content of the dough. The consumer prefers well-cooked and well-puffed *poori* with minimum oil pickup. Response surface methodology was used to optimize the processing conditions. The results of these studies showed that the extrusion pressure ranging from 3 to 6×10^5 Pa for the whole wheat flour dough with added moisture of 56 to 60 per cent was found to deliver the *poori* dough sheet at a uniform rate of extrusion. The optimum thickness of the *poori* dough sheet was found to be 2.0 mm. The frying time for the extruded dough sheet was observed in the range of 35 to 40 s with the temperature of the frying medium being in the range of 180 to185 °C. The oil uptake during frying was about 12±1 per cent and the textural shear force was found to be 9.9 N with an overall sensory score of 7.2±0.5 on nine point hedonic scale.

Poori Quality Improvement

Sudha and Venkateshwara Rao (2009) studied the effect of **hydoxypropylmethyl cellulose (HPMC)** on the rheological characteristics of whole wheat flour dough and *poori* making quality. Incorporation of HPMC at 0.5 and 1.0 per cent level increased the water absorption and dough stability whereas the resistance to extension and extensibility decreased. There was marginal increase in moisture and fat contents of *poori*. Addition of 0.5 per cent of HPMC increased oil uptake by about 0.5 per cent and improved the sensory quality of *poori*. Increased moisture retention upon addition of HPMC resulted in highly pliable and soft-textured *poori*. The *pooris* were softer and pliable even after 8 h of storage. The microscopic observations in *poori* showed partial starch gelatinization in the control samples, whereas HPMC containing *poori* showed higher gelatinization of starch.

Parimala and Sudha (2012) studied the effect of several **hydrocolloids** such as guar gum (GG), arabic gum (AG), carrageenan (CG), locust bean gum (LBG), xanthan gum (XN), hydroxypropyl methylcellulose (HPMC) and carboxymethylcellulose (CMC) at 0.5% w/w level on rheological and quality parameters of *poori* from whole wheat flour. The hydrocolloids like CMC, XN and HPMC increased the water absorption of *poori* dough, while it decreased in the case of other hydrocolloids. The dough development time and mixing tolerance index values increased, while dough stability was not affected. On the addition of hydrocolloids, there was a reduction in the pasting temperature, while the peak viscosity, hot paste viscosity and cold paste viscosity values increased. Hardness, cohesiveness and adhesiveness properties of the *poori* dough increased with the addition of hydrocolloids. All the hydrocolloids used, in general, helped in the retention of moisture in the *poori and* hence the *poori* remained softer and pliable, while there was a significant reduction in the oil content of *poori* samples containing hydrocolloids. Among the different hydrocolloids used, addition of guar gum at 0.5% (w/w) level led to *poori* having improved quality characteristics to a greater extent with respect to moisture retention, lowering of oil content upon frying with softer and pliable texture and better keeping quality.

Yadav and Rajan (2012) studied the effect of incorporation of wheat and oat brans in *pooris* to not only increase the dietary fibre content but also reduce oil uptake during deep fat frying. The *pooris* were prepared from the dough consisting of whole wheat flour, water (70 to 78 per cent based on mixograph water absorption), 2.0 per cent common salt (flour basis), and different levels of wheat bran (3 to 9 per cent, flour basis) and oat bran (6 to 12 per cent, flour basis). Response surface methodology was used to determine the optimum levels of wheat and oat brans to predict the levels of moisture and oil contents as well as overall acceptability of *pooris*. The results showed that the oat bran had significant positive effect on moisture retention and negative effect on oil uptake, whereas the wheat bran exhibited reverse effects. Based on compromised optimization, it was recommended to incorporate 3.0 g wheat bran and 11.0 g oat bran per 100 g flour for making dietary fibre rich *poori* with 20 per cent less oil than the control, and optimum acceptability (88.6).

Nutrient composition of *Poori*

Sharavathy et al (2001) while studying the *in vitro* measurement of nutritionally important starch fractions in some selected cereal based preparations reported the nutrient composition of *poori* as shown in **Table 8.**

Table 8: Nutrient composition of *poori* (g/100g dry basis).

Nutrient	Value
Protein (Nx6.25)	11.6
Fat	24.6
Free sugars	3.9
Starch	40.3
Soluble dietary fibre	1.4
Insoluble dietary fibre	6.4

Source: Shravathy et al (2001).

Poori was reported to provide about 18.2 per cent total starch with values for rapidly digestible starch, slowly digestible starch and resistant starch respectively being 2.6, 15.0 and 0.6 g per 100g fresh basis. The rapidly available glucose content of *poori* was 6.7g per 100g fresh basis. Addition of an accompaniment decreased the starch digestibility index which may be attributed to the slow and incomplete digestion of starch in *poori* as starch granules may be encapsulated by the fat absorbed. The availability of other nutrients such as zinc, calcium and iron from *poori* was found to be less when compared with chapatti (Grewal et al, 1999). Though *poori* has the highest values for true digestibility and net protein utilization, its biological value is lower when compared with *tandoori roti*, *paratha* and chapatti. The protein efficiency ratio of *pooris* when compared with chapattis, was found to decrease significantly by 22 per cent (Shyamala and Kennedy, 1962).

The nutritive value of *poori* may be improved through supplementation with other cereals or legumes, etc. Punia et al (2004) prepared *pooris* by incorporating amaranth (*Amaranthus tricolor*) leaves to increase the availability of calcium, iron and β-carotene in the daily diet.

Shelf-life of *Poori*

Since *poori* is deep fat fried, it picks up large amount of oil depending on the moisture content of the *poori* dough. Higher the moisture content of the dough, greater the oil pickup by the *poori* but concomitant decrease in moisture content of the *poori* (Shurpalekar and Shukla, 1992). However, compared to baked chapatti, *poori* has better keeping quality as well as acceptability because of its lower moisture content and slower staling changes.

ROOMALI ROTI

Introduction

Roomali roti is the larger version of chapatti. The name *roomali* or *rumali roti* is derived from the word *rumal* meaning handkerchief and *roti* meaning flat bread in Hindustani dialect. It is ultra-thin (~1mm) like a tissue paper with a diameter of 30 to 40 cm or larger, and is served folded like a handkerchief, hence the name *roomali roti*. It is popular in the Indian subcontinent particularly in India and Pakistan. In the Punjab Pakistan it is also known as *lamboo roti* simply meaning long in the Punjabi dialect. It can be as large in size as a bath towel and baked on large inverted concave shaped iron griddle. It is usually prepared from a combination of whole wheat flour (*Atta*) and all-purpose flour (*maida*), the proportion varying from preparation to preparation. The flour mix may or may not be supplemented with baking powder and edible oil but invariably contains common salt to taste.

Traditional Method of Preparation

Atta, maida, common salt and baking powder (if used) are mixed well by sieving them together twice. The oil is then rubbed into the flour mix. Water is added gradually, mixed and kneaded well to make smooth and elastic dough. The dough is then set aside covered with a moist muslin cloth for 30 min. The dough is divided into small balls (40 to 50 g). The dough ball is then smeared with little dry flour and rolled like a thin (~1mm) chapatti of about 30 to 40 cm diameter circle. The dough circle should be very thin at the edges. An experienced cook may sheet the dough after applying sufficient dry flour into chapatti like sheet by slapping between the palms of hands and repeated tossing up in the air and stretching simultaneously with the help of reverse of palms of both hands. The dough sheet is then flipped carefully over the preheated large inverted griddle (*tawa*) and baked till small bubbles start appearing over the surface and the underside is light brown. It is then turned upside down at once and baked till the other side is slightly brown too. It is then removed from the griddle and folded like a handkerchief to facilitate serving. Alternatively, it may be rolled like a Swiss roll to arrest drying during storage. To prevent sticking, the griddle should be really very hot and sprinkled with salted water. Instead of *tawa*, inverted *kadahi* (Indian wok) may be used to bake the *roomali roti.*

References

AACC (2000). *Approved methods of the American Association of Cereal Chemists,* 10thEd. The Association, St, Paul, MN, USA.

Abrol, YP (1971). Studies on chapatti quality. Presented at seminar "*Wheat in Indian Diets*", pp 30-31. **Cross reference.**

Abrol YP (1972). Studies on chapatti quality. I. Discoloration of chapattis. *Bull. Grain Tech.* **10:** 41- 43.

Abrol YP and Mehdi V (1971). On the sweetness of whole wheat meal. *Current Sci.* **40:** 381- 383.

Abrol YP, Mehdi V, Uprety DC, Tikoo S and Ram A (1972a). Studies on chapatti quality. III. Preservation of discoloration of chapaties. *Bull. Grain Technol.* **10:** 100- 106.

Abrol YP, Ram A and Mehdi V (1972b). Studies on chapatti quality. II. Role of sugars, amino acids and phenolics in discolouration of chapattis. *Bull. Grain Technol.* **10:** 3- 10.

Abrol YP, Uprety DC and Ram A (1971).Browning of chapaties. *J Food Sci. Technol.* **8:** 26- 27.

Achaya KT (1994). *Indian Food: A historical companion*. Oxford University Press, England.

Acton JC, Breyer L and Satterlee LD (1982). Effect of dietary fibre on the in vitro digestibility of casein. *J. Food Sci.* **47:** 556- 560.

Adsule RN, Lawande KM, Dhotre VA and Kadam SS (1985). Studies on some promising cultivars of wheat. II. Quality of Atta and Chapati. *Bull. Grain Technol.* **23:** 134 – 138.

Agte VV and Joshi SR (1997). Effect of traditional food processing on phytate degradation in wheat and millets. *J. Agric. Food Chem.* **45:** 1659- 1661.

Ahluwalia P and Kaur A (2001).Studies on use of *isabgol* (*Psyllium mucilloid*) husk in *atta* for chapatti making. *J. Food Sc. Technol.* **38:** 75- 78.

Ahmed A, Anjum FM, Rehman S, Randhawa MA and Farooq U (2008). Bioavailability of calcium, iron and zinc fortified whole wheat flour chapatti. *Plant Foods Human Nutr.***63:** 7- 13.

Ahmed A, Anjum FM, Ahmad A, Khalid N, Randhawa MA, Ahmed Z and Farid W (2013). Effects of hydrocolloids on partial baking and frozen storage of wheat flour chapatti. *Food Sci. Technol.* Res. **19:** 97- 103.

Ajmera S, Gordon J and Janus D (2015). System and method for preparing *naan* bread. *US Patent* No. US 90445021 B2.WWW.Googlede / patents / US 90445021.

Akhtar S, Anjum FM, Rehman SU and Sheikh MA (2009). Effect of mineral fortification on rheological properties of whole wheat flour. *J. Texture Studies* **40:** 51- 65.

Althwab S, Carr TP, Weller CL, Dweikat IM and Schlogel V (2015). Advances in grain sorghum and its co-products as a human health promoting dietary system. *Food Research Int'l.* **77:** 349- 359. **https://doi.org/10.1016/j.foodres.2015.08.011.**

Amerine MA, Pangborn RM and Roessler EB (1965). *Principles of sensory evaluation of food*, Academic Press, New York.

Anderson, EN. (2005). *Everyone eats: understanding food and culture*. New York: New York University Press. pp. 47–48.

Anglani C (1998). Sorghum for human food- A review, *Plant Foods Human Nutr.* **52:** 85- 95.

Anjum FM, Ahmad I, Butt MS, Sheikh MA and Pasha I (2005). Amino acid composition of spring wheats and losses of lysine during chapati baking. *J. Food Composition Analysis* **18:** 523- 532.

Anjum FM, Ali A and Chaudhry NM (1991). Fatty acids, mineral composition and functional (bread and chapatti) properties of high protein and high lysine barley lines. *J. Sci. Food Agric.***55:** 511- 519.

Anjum FM, Pasha I, Ahmad S, Khan M I and Iqbal Z (2008). Effect of emulsifiers on wheat-potato composite flour for the production of leavened flat bread (naan). *Nutri. Food Sci.* **38:** 482- 491.

Anjum MF, Lookhart GL, Walker CE (2000). High-molecular-weight glutenin subunit composition of Pakistani hard white spring wheats grown at three locations for 2 years and its relationship with end-use quality characteristics. *J. Sci. Food Agric.* **80:** 219- 225.

Anon. (1976). PL- 480.*Annual Research Report of Project, Development and Evaluation of Triticale*. Dept. of Plant Breading, Punjab AgriculturalUniversity, Ludhiana (India).

AOAC (2000). *Official methods of analysis*. Association of Official Analytical Chemists, 16th ed. Washington, USA.

Arthur JR (1999).Functional indicators of iodine and selenium status. *Proc. Nutr. Soc.* **58:** 507- 512.

Arya SS, Ananthanarayan L, Rodrigues L and Waghmare A (2016). Fractionation and reconstitution of whole wheat flour and its effect on dough and chapatti quality. *J. Food Measur.* **10:** 614- 624.

Arya SS and Sonawane SK (2016). Impact of fiber mixture on dough and chapatti quality using D-optimal response surface methodology. *J. Microbiol. Biotech. Food Sci.* **5:** 424- 433, **doi:10.15414/jambfs.2016.5.5.424-433.**

Arya SS (1984). Traditional Indian foods: some recent developments. *Def. Sci. J.* **42:** 173- 182.

Arya SS, Mohan MS and Nath H (1971). Studies on packaging and storage of *atta* (wheat flour) under tropical conditions. *J. Food Sci. Technol.* **8:**134- 139.

Arya SS and Murthy MCN (1983). Effect of surfactant, fatty acids and glycerides on the gelatinization viscosity of atta (wheat flour). *J Food Sci. Technol.* **20:** 116- 117.

Arya SS, Premavalli KS and Parihar DB (1976). Technical note: precursors of carbonyls of chapaties. *J Food Technol.***11:** 543- 547.

Arya SS, Vanaja N and Parihar DB (1978). Role of lipids in chapatti baking. In: Fats and oils in relation to food products and their preparation. *Proceedings of Fats and Oil Symposium*, AFST (I) and Defence Food Research Lab., Mysore, India, pp 146- 148.

Arya SS, Vidyasagar K and Parihar DB (1977). Preservation of chapaties. *Lebens.Wiss. Technol.* **10:** 208- 210.

Asif M, Rooney LW, Ali R and Riaz MN (2013). Applications and opportunities of pulses in food systems: A review. *Crit. Rev. Food Sci. Nutr.* **53:** 1168- 1179.

Austin A (1978a). Improvement of wheat quality. *Indian Farming* **27:** 23-25, 27.

Austin A (1978b). *Wheat Research in India*, Indian Council of Agriculture Research, New Delhi, India, p.188.

Austin A and Mirchandani RL (1971). A comparative study of the sugar content and diastatic activity in whole wheat meal and flour of improved Indian and Mexican wheat varieties. *Bull. Grain Technol.* **9:** 3- 6.

Austin, A., Ram, A. (1971). Studies on chapati-making quality of Wheat, Indian Agric. Res. Inst., New Delhi, India. *I.C.A.R. Technical Bulletin* No. 31.

Axford DWE, McDermott EE and Redman DG (1979). Note on the sodium dodecyl sulfate test of bread making quality, comparison with Pelshenke and Zeleny tests. *Cereal Chem.* **56:** 582- 584.

Aziz MA (1960). *Cereals and Pulses*, Publ. Department of Agriculture, West Pakistan. **Cross reference.**

Aziz MA and Bhatti HM (1960). Quality considerations for chapattis (unleavened pancakes). *Agriculture (Pakistan)* **13:** 157 - 163. **Cross reference.**

Badgujar JB, Ghodke SK, Ananthanarayan L and Rodrigues L (2010). Preservation of chapatti (Indian unleavened flat bread) using retort processing. Presentation Abstract, *Institute of Food Technologists, Chicago, USA*, 2010.

Baik MY and Chinachoti P (2000). Moisture redistribution and phase transitions during bread staling. *Cereal Chem.* **77:** 484- 488.

Bains GS (1949). Effect of commercial fertilizers and green manure on yield and nutritive value of wheat. I. Nutritive value with respect to total phosphorus, phytic phosphorus, non-phytic phosphorus and calcium content of the grain. *Cereal Chem.* **26:** 317. **Cross reference.**

Bains GS and Irvine GN (1965). Studies on the quality of some improved varieties of Indian wheats. *J. Sci. Food Agric.* **16:** 526- 535.

Bains GS, Reddy SK, Bhatia DS and Subrahmanyan V (1952). Investigation on the utilization of tapioca flour for chapaties, *pooris* and vermicelli. *Bull. Cent. Food Technol. Res. Inst., Mysore* **2:** 38- 42.

Bajaj M, Kaur A and Sidhu JS (1990). Studies on debranning of Punjab wheats. II. Effect on rheological characteristics and chapati making quality. *J. Plant Sci. Res.* **6:** 40-43.

Bajwa U (1997). Soy enriched flat bread (chapaties): evaluation of texture using Instron Universal Testing Machine. *Adv. Food Sci.* **19:** 167- 171.

Bakhshi, AK, Sekhon, KS, Sehgal, KL and Gill KS (1979). Chapati-making performance of some promising wheat strains. *J. Res, PAU.* **16:** 267-271.

Bakshi AK, Sehgal KL, Singh RP and Gill KS (1989). Effect of bread wheat, durum wheat and triticale blends on chapatti, bread and biscuit. *J. Food Sci. Technol.* **26:** 191- 193.

Banecki H (1982). Gluten- the deciding factor in staling of bread. *Gert.Mehl Brot.* **36:** 272. **Cross Reference.**

Banerji A, Ananthnarayan L and Lele S (May 2017). Rheological and nutritional studies of amaranth enriched wheat chapatti (Indian flat bread). *J. Food Process. Preserv.* DOI**: 10.1111/jfpp.13361.**

Bano R, Hussain S, Li J, Ali S, Jin W, Li H, Ali A and Hussain M (2014). Nutritional improvement of whole wheat flour chapatti by supplementation of tartary buckwheat flour. *Food Sci. Quality Management* **34:** 78- 85.

Bansal G, Singh R P and Nanda G S (2000). Suitability of advanced wheat genotypes for bread, biscuit and *chapaties. J. Res. PAU.***47:** 63-70.

Bass J and Caul JF (1972). Laboratory evaluation of three protein sources for use in chapatti flour. *J. Food Sci.* **37:** 100- 104.

Bass JE (1988). Wheat flour milling. In *Wheat: chemistry and technology*, Pomerantz Y (Ed). American Association of Cereal Chemists, St. Paul, MN.

Bechtel WG, Meisner DF and Bradley WB (1953). The effect of the crust on the staling of bread. *Cereal Chem.* **30:** 160- 168.

Bedekar, BR (2001). Chapati- Indian Bakery Product of the 21st Century, *Indian Food Industry* **20:** 21-23, 63.

Bender AE (1972). Processing damage to protein food (Review). *J. Food Technol.* **7:** 239- 250.

Berry SK (1973). Volatile constituents of model systems related to bread. *Ph. D. Thesis*, the University of Leeds, Leeds, England.

Berry SK and Gramshaw JW (1988). Influence of starch plus gluten on the non-enzymatic browning reaction of glucose-glutamic acid system. *J. Agric. Food Chem.* **36:** 1265- 1267.

Bhargawa S, Ranote PS, Kapoor S and Sharma S (2014). Quality evaluation of aloe (*Aloe barbedensis*, Miller) juice supplemented chapatti. *Asian Resonance* III: 1- 6.

Bhargava S, Jyothi Lakhsmi A and Indrani D (2012). Effect of barley, banana and soy protein isolate on rheological, microstructural and nutritional characteristics of North Indian parotta. *J. Texture Studies* **43:** 246- 256.

Bhat CM and Vivian VM (1980). Effect of incorporation of soy, peanut and cottonseed flours on the acceptability and protein quality of chapaties. *J. Food Sci. Technol.* **17:** 168- 171.

Bhatnagar T, Suchdev A and Johari RP (2002). Molecular characterization of glutenins in wheat varieties differing in chapatti quality characteristics. *J. Plant Biochem. Biotechnol.* **11:** 33- 36.

Bhatty RS (1995). Laboratory and pilot plant extraction and purification of β-glucans from hull-less barley and oat bran. *J. Cereal Sci.* **22:** 163- 170.

Bhavyashree SH, Prakash J, Platel K and Srinivasan K (2009). Bioaccessibility of minerals from cereal-based composite meals and ready-to-eat foods. *J. Food Sci. Technol.* **46:** 431- 435.

Bhise S and Kaur A (2013). Development of functional chapatti from texturized deoiled cake of sunflower, soybean and flax seed. *Int'l. J. Eng. Res. Applic.* **3:** 1581-1587.

Bhoir SA, Mupalla SR, Kanatt SR, Chawla SP and Sharma A (2015). Radapperization of ready- to-eat shelf-stable traditional Indian bread- *Methi Paratha*. *Radiation Physics and Chemistry* **111:** 24- 27.

Bora P (2014). Anti-nutritional factors in foods and their effects. J. *Acad. Inds. Res. (JAIR).* **3:** 285- 290.

Bourne MC (2002). *Food Texture and Viscosity- concept and measurement*, 2nd Edition. Publisher: Elsevier Science and Technology.

Chakrabarti-Bell S, Wang S, Patel MJ, Weiss RM and Austin PJ (2013). Bubbles in chapatti dough. *J. Cereal Sci.* **57:** 504- 513.

Chari ST and Sreenivasan A (1980). Protein concentrates from shark. *Fishery Technol.* **17**: 115- 117.

Chatelanat RP (Ed.) (1973). *Composite Flour Programme*, FAO, Rome, Italy.**Cross reference.**

Chaudhary B and Mukhopadhyay K (2012). *Syzygium cumini* (L.) Skeels: Apotential source of neutraceuticals. *Int'l. J. Pharm. Biol. Sci.* **2:** 46- 53.

Chaudhri AB and Muller HG (1970). Chapaties and chapati flour. *Milling* **152:** 22, 24-25.

Chaudhry MS (1968). Preparation and evaluation of *attas* and chapatis from US wheats. *Dissertation Abstracts, Section B.* **29** (1): 236.**Cross reference**.

Chaudhry N, Dangi P and Khatkar BS (2016). Assessment of molecular weight distribution of wheat gluten proteins for chapatti quality. *Food Chem.* **199:** 28- 35.

Chaudri AB and Muller HG (1972). The destruction of thiamine during chapati baking. *J.Food Technol.* **9:** 123- 124.

Chauhan A, Saxena DC and Singh S (2016). Physical, textural and sensory characteristics of wheat and amaranth flour blend cookies. *Cogent Food Agric.* **2**(10)**:DOI: 10.1080/23311932.2015.1125273.**

Chauhan GS and Santosh Kumari (1990). Evaluation of wheat-defatted soy flour blends for chapatti making quality. *Beverage and Food World.* **17:** 40- 42.

Chawla VK and Kapoor AC (1982).Chemical composition and protein quality of wheat-triticale chapattis. *J. Food Sci.* **47:** 2015- 2017.

Chopra K and Bhat CM (1975). Acceptability of chapaties prepared from high yielding varieties of wheat. *Ind. J. Nutr. Diet.* **12:** 120- 123.

Chowdhury AR, Bhattacharyya AK and Chattopadhyay P (2012). Study on functional properties of raw and blended jackfruit seed flour (a non-conventional source) for food applications. *Indian J. Natural Products Resources* **3:** 347- 353.

Christian P and Seshadri S (1989). Counteracting the inhibitory effect of tea on the in *vitro* availability of iron from cereal meals. *J. Sci. Food Agric.* **49:** 431- 436.

Dandy DAV (1992). Composite flour- past, present and future- A review with special emphasis on the place of composite flour in the semi-arid zones. In *Utilization of sorghum and millets* (Gomez ML, House LR, Rooney LW and Dandy DAV, Eds.). International Crop Research Institute for the Semi-arid Tropics, Patancheru, Andhra Pradesh, India, pp 67- 73.

Dar BN, Sharma S, Singh B and Kaur G (2014). Quality assessment and physicochemical characteristics of bran enriched chapattis. *Int'l. J. Food Sci.* **2014:Article ID 689729, 6 pages. http://dx.doi.org//10.1155/2014/689729.**

Darak Priya J and Pawar V (2015). Utilization of fenugreek (*Trigonella foenum-graecum* L.) malt in unleavened flat bread (chapatti). *Beverage and Food World* **42:** 29-32, 35.

Das A, Thorat SS, Kulkarni SD and Ingle UM (1990). Comparative evaluation of new wheat cultivars for physicochemical and chapatti making quality. *J.Maharashtra Agricultural Universities***15:** 83- 89.

Das BK, Sharma A, Marathe SA, Rao VS and Bhagwat SG (2006). Evaluation of bread and chapatti making quality of wheat selections and parents differing in HMW subunits of glutenin. *J. Food Sci. Technol.* **43:** 69- 72.

De Reuter D (1978). Composite flours. In: *Advances in Cereal Science and Technology*, Pomeranz Y (Edit.), pp 349- 385. St. Paul, USA, AACC.

Deshpande SS,Mishra A and Mishra M (2001). Preparation and organoleptic evaluation of soy-blended food products. *J. Food Sci. Technol.* **38:** 291-293.

Devi PB, Vijayabharathi R, Sathyabama S, Malleshi NG and Priyadarisini VB (2014). Health benefits of finger millet (*Eleusine coracana*, L.) polyphenols and dietary fibre- a review. *J. Food Sci. Technol.***51:** 1021- 1040.

Dexter JE and Matsuo RR (1978). The effect of gluten protein fractions on pasta dough rheology and spaghetti-making quality. *Cereal Chem.* **55:** 44- 58.

Dhaliwal YS, Hatcher DW, Sekhon KS and Kruger JE (1996).Methodology for the preparation and testing of chapattis produced from different classes of Canadian wheat. *Food Res. Int'l.* **29:** 163-168.

Dhingra S, Mehta U and Bhat CM (1982). Chapati making qualities of pre-released high yielding varieties of wheat. *Bull. Grain Technol.* **20:** 164 – 167.

Dogra J, Dhaliwal YS and Kalia M (2001). Effect of soaking, germination, heating and roasting on the chemical composition and nutritional quality of soybean and its utilization in various Indian leavened products. *J. Food Sci. Technol.* **38:** 453-457.

Dubois DK and Hoover WJ (1981). Soya protein products in cereal grain foods. *J. Am. Oil Chemists Soc.* **58:** 343- 346.

Duss R and Nyberg L (2004). Oat soluble fibers (β-glucans) as a source of healthy snack and breakfast foods. *Cereal Foods World* **49:** 320- 325.

Dutta SK, Russell RM and Chowdhury B (1980). Folate content of North Indian breads.*Nutr. Rep. Int'l.* **21:** 251- 156.

DWR Progress Report (2008). Wheat quality. *All India Coordinated Wheat and Barley Improvement Programme,* Eds. R.K.Gupta, D. Mohan, Sewa Ram and B. Mishra, Vol. 4. Directrate of Wheat Research, Karnal, India.

Ebeler SE and Walker CE (1983). Wheat and composite flour chapaties; effects of soy flour and sucrose-ester emulsifiers. *Cereal Chem.* **60:** 270- 275.

Eggum BO and Duggal SK (1977). The protein quality of some Indian dishes prepared from wheat. *J. Sci. Food Agric.* **28:** 1052- 1056.

Elwood PC, Benjamin IT, Fry FA, Eakins JD, Brown DA, De Kock PC and Shah JU (1970). Absorption of iron from chapatti made from wheat flour. *Am. J. Clin. Nutr.* **23:** 1267- 1271.

Ezekiel R and Singh N (2011). Use of potato flour in bread and flat bread. In*: Flour and Breads and their Fortification in Health and Disease Prevention*, Preedy VR, Watson RR and Patel VB (Eds.), Chapter 23, pp 247- 259. Elsevier/ Academic Press, Amsterdam; Boston, 2011.

Fabjan N, Rode J, Kosir IJ, Zang Z and Kreft I (2003). Tartary buckwheat as a source of dietary rutin and quercetin. *J. Agri. Food Chem.* **51:** 6452- 6455.

Fadda C, Sanguinetti AM, Del Caro A, Collar C and Piga A (2014). Bread staling: updating the view. *Compr. Rev. Food Sci. Food Saf.* **13:** 473- 492.

FAO/WHO (1991). Protein quality evaluation:*Report of a joint FAO/WHO expert consultation,* Food and Agriculture Organization of the United Nations, Rome.

FAO / WHO (2004). Vitamin and mineral requirements in human nutrition (second edition): *Report of expert consultation*, Geneva, Switzerland.

Faridi H (1988). Flat breads. In *Wheat: Chemistry and Technology*, Vol. II. Pomeranz Y (Ed.), American Association of Cereal Chemists, St. Paul, MN.

Farook Z, Rehman S and Abid M (2013). Application of response surface methodology to optimize composite flour for the production and enhanced storability of leavened flat bread (*naan*). *J. Food Process. Preserv.* **37:** 939- 945.

Farrand EA (1972). Controlled levels of starch damage in commercial United Kingdom bread flour and effects on absorption, sedimentation value, and loaf quality. *Cereal Chem.* **49:** 479- 488.

FDA (1998). Food labeling. Health claims: Soluble fibre from certain foods and risk of coronary heart disease. *21 CFR ©101, 81. Code of Federal Regulations*, Washington DC.**Cross reference**

Food Safety and Standards Act, 2006. Food Safety and Standards Authority of India, Ministry of Health and Family Welfare, Govt. of India, New Delhi, 2011.

Forman TM and Evanson DN (2010). Antistaling process for flat bread. *Patent Publication No. WO 2010124206 A1*.WWW.google.com/patents /WO 2010124206 A 1.

Gandhi AP and Bourne MC (1988).Technical note: Effect of added soybean paste and instant soydhal on toughness and rate of staling of chapatis. *Int'l. J. Food Sci. Technol.* **23:** 411- 414.

Gandhi AP, Kotwaliwale N, Taimini V, Kawalkar J, Srivastav DC, Parihar VS and Raghunadh P (2000). Effect of incorporation of defatted soyflour on the texture of chapattis *.J. Texture Studies* **31:** 609- 614.

Gandhi AP, Mishra VK and Ali N (1983). Organoleptic assessment of full fat soyflour in various indigenous products. *J. Food Technol.* **18:** 771-775.

Ghavidel RA and Prakash J (2006). Effect of germination and dehulling on functional properties of legume flours. *J. Sci. Food Agric.* **86:** 1189- 1195.

Ghodke SK (2009). Effect of guar gum on dough stickiness and staling of chapatti- an Indian unleavened flat bread. *Int'l. J. Food Eng.* **5**: 1- 19.

Ghodke SK and Ananthanarayan L (2007). Influence of additives on rheological characteristics of whole-wheat dough and quality of chapatti (Indian unleavened flat bread).Part I-hydrocolloids. *Food Hydrocoll.* **21:**110-117

Ghodke SK, Ananthanarayan L and Rodrigues L (2009). Use of response surface methodology to investigate the effects of milling conditions on damaged starch, dough stickiness and chapatti quality. *Food Chem.* **112:** 1010- 1015.

Ghosh KG, Srivatsava AN, Nirmala N and Sharma TR (1973). Development and application of fungicide wrappers in food preservation- Wrappers obtained by impregnation method. *J. Food Sci. Technol.* **10:** 105- 109.

Ghosh KG, Srivatsava AN, Nirmala N and Sharma TR (1977). Development and application of fungistatic wrappers in food preservation. Part II. Wrappers made by coating process. *J. Food Sci. Technol.* **14:** 261- 264.

Ghufran Saeed SMP, Saqib A, Mubarak A, Rashida A and Fred S (2009). Influence of rice bran on rheological properties of dough and in the new product development. *J. Food Sci. Technol.* **46:** 62- 65.

Giese J (2003). Texture measurement in foods. *Food Technol.* **54:** 62- 65.

Gill BS, Sodhi NS and Kaur M (2005). Effect of chickpea, ghee, sodium chloride, mixing time and resting time on instrumental texture and sensory quality of chapatti. *J. Food Sci. Technol.* **42:** 481- 488.

Gill KS, Sekhon KS and Nanda GS (1980). Development of high yielding wheat varieties superior in bread, biscuit and chapatti making properties. *Proc. Seminar "Wheat Your Need", New Delhi*, p 189. **Cross reference**.

Goel S, Rathore M, Grewal S, Jain N, Singh BK, Ahlawat AK, Singh AM, Singh PK and Singh NK (2015). Effect of allelic variations in triticin on bread-and chapatti-making qualities of wheat (*Triticum aestivum*). *Agric Res.* **4**(2):139–151. **DOI 10.1007/s40003-015-0150-1**

Gomez M, Ronda F, Blano CA, Caballero PA and Apestegui A (2003). Effect of dietary fibre on dough rheology and bread quality. *European Food Res. Technol.* **216:** 51- 56.

Gray JA, Bemiller JN (2003). Bread Staling: Molecular Basis and Control. *Compr. Rev Food Sci. Food Saf.* **2:** 1- 21.

Greer EN and Stewart BA (1959). The water absorption of wheat flour. *J. Sci. Food Agric.* **10:** 248- 252.

Grewal HK and Hira CK (2003). Effect of processing and cooking on zinc availability from wheat (*Triticum aestivum*). *Plant Foods Human Nutr.* **58**(3)**:** 1- 8.

Grewal HK, Hira CK and Kawatra BL (1999). Zinc, calcium and iron availability using molar ratios in processed and cooked wheat products. *J. Food Sci. Technol.* **36:** 453- 456.

Grewal HK, Hira CK and Kawatra BL (2000). Iron availability from processed and cooked wheat products using hemoglobin regeneration efficiency method. *Nahrung* **44:** 398- 402. **FSTA abstract.**

Grover K, Sadana BK and Hira CK (1994). Availability of trace minerals from cereal preparations. *Indian Food Packer* **48:** 39- 44.

Gujral HS and Gaur S (2002). Effect of barley flour, wet gluten and liquid shortening on the texture and storage characteristics of chapatti. *J. Texture Studies* **33:** 461- 469.

Gujral HS and Gaur S (2005). Instrumental texture of chapatti as affected by barley flour, glycerol monostearate and sodium chloride. *Int'l. J. Food Properties* **8:** 377- 385.

Gujral HS and Pathak A (2002). Effect of composite flours and additives on the texture of chapatti. *J Food Eng.* **55:** 173- 179.

Gujral HS, Sharma P, Bajaj R and Solah V (2012). Effects of incorporating germinated brown rice on the antioxidant properties of wheat flour chapatti, Food Sci. Technol. Int'l. **18:** 47 – 54. **DOI:10.1177/1082013211414173.**

Gujral HS, Sharma P, Gill BS and Kaur S (2013). Effect of incorporating hydrothermal, kilned and defatted oats on antioxidant and chapatti making properties of wheat flour. *Food Chem.* **138:** 1400- 1406.

Gujral HS, Singh GS and Rosell CM (2008). Extending shelf-life of chapatti by partial baking and frozen storage. *J. Food Eng.* **89:** 466- 471.

Gupta A, Singh C, Kumar V, Tyagi BS, Tiwari V, Chatrath R and Singh GP (2018). *Wheat varieties notified in India since 1965*. ICAR- Indian Institute of Wheat & Barley Research, Karnal-1321001, India: 101 pp.

Gupta S and Kawatra BL (1992). Effect of cereal-legume chapatti diets on absorption and retention of calcium. *Plant Foods Human Nutr.* **42:** 165- 173.

Gupta TR (2001). Individual heat transfer modes during contact baking of Indian unleavened flat bread *(chapati)* in a continuous oven. *J. Food Eng.* **47:** 313- 319.

Gurushree MN, Nandini CR, Pratheeksha K and Prabhasankar P (2011). Design, development and performance evaluation of chapatti press cum vermicelli extruder. *J. Food Sci. Technol.* **48:** 218- 224.

Haq MYI and Chaudhry MS (1976). Supplementation of *roti* and *naan* with chickpea flour. *Pak. J. Sci. Ind. Res.***19:** 66- 71.

Harinder K, Kaur B and Sharma S (1999). Studies on the baking properties of wheat: pigeon pea flour blends. *Plant Foods Human Nutr.* **54**: 217- 226.

Harland BF (1989). Dietary fibre and mineral bioavailability. *Nutr. Res. Rev.* **2:** 133- 147.

Hashmi I and Wootton M (1995). *Tandoori* bread production and quality. *Food Australia* **47:** 366- 368. **FSTA Abstracts.**

Hatcher DW, Kruger JE and Dhaliwal YS (1997). Effect of roller mill extraction rate on the chapatti – making quality of Canadian flours. *Cereal Chem.* **74** (4): 502- 508.

Hemalatha MS, Bhagwat SG, Salimath PV and Prasada Rao UJS (2012). Enhancement of soluble dietary fibre, polyphenols and antioxidant properties of chapattis prepared from whole wheat flour dough treated with amylases and xylanase. *J. Sci. Food Agric.* **92:** 764- 771.

Hemalatha MS, Leelavathi K, Salimath PV and Prasada Rao UJS (2014). Control of chapatti staling upon treatment of dough with amylases and xylanase. *Food Bioscience* **5:** 73- 84.

Hemalatha MS, Manu BT, Bhagwat SG, Leelavathi K, and Prasada Rao UJS (2007a). Protein characteristics and peroxidase activities of different Indian wheat varieties and their relationship to chapatti-making quality. *Eur. Food Res. Technol.* **225:** 463- 467.

Hemalatha MS, Prasada Rao UJS, Leelavathi K and Salimath PV (2010). Influence of amylases and xylanase on chemical, amylograph properties and microstructure of chapatti. *LWT- Food Sci. Technol.***43:**1394- 1402.

Hemalatha MS, Sai Manohar R, Salimath PV and Prasada Rao UJS (2013). Effect of added arabinoxylans isolated from good and poor chapatti making wheat varieties on rheological properties of dough and chapatti making quality. *Food Nutr. Sci.* **4**: 884- 892.

Hemalatha S, Platel K and Srinivasan K (2007b). Influence of heat processing on the bioaccessibility of zinc from cereals and pulses consumed in India. *J. Trace Elem. Med. Biol.* **21:** 1- 7.

Hemdane S, Jacobs PJ, Dornez E, Verspreet J, Delcour JA, and Courtin CM (2016). Wheat (*T*riticum aestivum L.) bran in bread making: A critical review. *Compre. Rev. Food Sci. Food Saf.* **15:** 28- 42.

Hira CK and Kaur AP (1993). Phytate/zinc and phytate x calcium/zinc ratios of common cereals, legumes and their combinations. *J. Food Sci. Technol.* **30:** 213- 215.

Hoffpauer DW (2005). New applications for whole rice bran. *Cereal Foods World* **50:** 173- 174.

Holiday R and Phillips K (2001). Health benefits of the sunflower kernel. *Cereal Foods World* **46:** 205- 208.

Honeytop Foods, UK (2000). *Naan* bread revolution. *Food Manufacture* **75** (2): 38- 39.

Hurrel RF (2002). Fortification: overcoming technical and practical barriers. *J. Nutr.* **132:** 806S- 812S.

Hussain S, Anjum FM, Butt MS, Alani MS and Shabbir MA (2012). Development and evaluation of nutritionally superior baked products containing flax seed. *Pakistan J. Nutr.* **11:** 160- 165.

Hussain M and Satti MH (1969). Development of long- life roti (bread). *Pak. J. Sci. Ind. Res.* **12:** 408- 411.

Inamdar AA and Prabhasankar P (2017). Influence of stone *chakki* settings on the characteristics of whole wheat flour (*atta*) and its chapati making quality. *J. Food Process. Preserv.* **41 (3):**12966. **DOi/10.1111/jfpp.12966.**

Inamdar AA, Sakhare SD and Prabhasankar P (2015). Chapati making quality of whole wheat flour (*atta*) obtained by various processing techniques. *J. Food Process. Preserv.* **39:** 3032- 3039.

Indira M and Kurup PA (2003). Black gram: A hypolipidemic pulse. *Natural Product Radiance.* **2:** 240- 242.

Indrani D (1998). Rheological characteristics of wheat flour dough in relation to quality of *parotta* – an Indian traditional food. *PhD Thesis*, Mysore University, Mysore, Karnataka State, India.

Indrani D, Jyothsana Rao S, Prabhasankar P and Venkateswara Rao G (2003). Chemical, rheological and *parotta*-making characteristics of flourmill streams. *Eur. Food Res. Technol.* **217:** 219- 223.

Indrani D, Jyothsana Rao S, Udaya Sankar K and Venkateswara Rao G (2000). Changes in the physical-chemical and organoleptic characteristics of *parotta* during storage. *Food Res. Int'l.* **33:** 323- 329.

Indrani D, Jyotsana R and Venkateswara Rao G (2010). Influence of fenugreek seed powder on the dough rheology, microstructure and quality of parotta- An Indian flat bread. *J. Texture Studies* **41:** 2o8- 2023.

Indrani D, Parbhasankar P, Jyotsna R and Venkateswara Rao G (2002). Effect of storage on the rheological and *parotta*-making characteristics of frozen *parotta* dough. *Eur. Food Res. Technol.* **215:** 484- 488.

Indrani D, Parbhasankar P, Jyotsna R and Venkateswara Rao G (2007).Influence of whey protein concentrate on the rheological characteristics of dough, microstructure and quality of unleavened flat bread (parotta). *Food Res. Int'l.* **40:** 1254- 1260.

Indrani D and Sai Manohar R (2007). Alveograph as a tool to assess the quality characteristics of wheat flour for *parotta* making. *J. Food Eng.***78:** 1202- 1206.

Indrani D, Swetha P, Soumaya C, Jyotsna R and Venkateswara Rao G (2011). Effect of multigrains on rheological, microstructural and quality characteristics of north Indian parotta. *LWT- Food Sci. Technol.* **44:** 719- 724.

Indrani D and Venkateswara Rao G (2000). Effect of chemical composition of wheat flour and functional properties of dough on the quality of South Indian *parotta*. *Food Res. Int'l.* **33:** 875- 881.

Indrani D and Venkateswara Rao G (2001). Optimization of the quality of South Indian *parotta* by modeling the ingredient composition using the response surface methodology. *Int'l. J. Food Sci. Technol.* **36:** 189- 197.

Indrani D and Venkateswara Rao G (2003). Influence of surfactants on rheological characteristics of dough and quality of *parotta*. *Int'l. J. Food Sci. Technol.* **38:** 47- 54.

Indrani D and Venkateswara Rao G (2004). Effect of processing conditions on the quality of South Indian *parotta*-An Indian traditional food. *J. Food Quality* **27:** 55- 72.

Indrani D and Venkateswara Rao G (2006). Effect of additives on rheological characteristics and quality of wheat *parotta*. *J. Texture Studies* **27:** 315- 338.

Indrani D and Venkateswara Rao G (2007). Rheological characteristics of wheat flour dough as influenced by ingredients of *parotta*. J. Food Eng. **79:** 100- 105.

ISI (1979). Specifications for chapati machine. *Indian Standard*: 9246, Indian Standards Institution, New Delhi, India.

Institute of Food Technologists, USA (1981). Sensory evaluation guide for testing food and beverage products. *Food Technol.* **35** (11)**:** 50- 59.

Islam Md N and Johansen HB (1987). Physicochemical tests – A bases for selecting the size of wheat flour. *J. Food Sci. Technol.* **24:** 136-138.

Izydorczyk MS and Biliadeis CG (1995). Cereal arabinoxylans: advances in structure and physicochemical properties. *Carbohydr. Polym.* **28:**33- 48.

Jackel SS (1987). More wheat with superior baking quality is needed. *Cereal Foods World* **32:** 323- 325.

Jha J, Bains NS and Singh RP (2010). Rheological and baking characteristics of a diverse set of advanced winter x spring lines of wheat. *J. Res. PAU.* **47:** 4- 13.

Jinshui W, Cristna MR and Carmen BB (2002). Effect of addition of different fibres on wheat dough performance and bread quality. *Food Chem.* **79:** 221- 226.

Jisha S, Padmaja G, Moorthy SN and Rajeshkumar K (2008). Pretreatment effect on the nutritional and functional properties of selected cassava-based composite flours. *Innovative Food Sci. Emerging Technologies* **9:** 587- 592. www/researchgate.net/publication/222645289_Pre.

Joshi P and Mathur B (2015). Development of value added products from the leaf powders of dehydrated less utilized green leafy vegetables. *Nutr. Food Sci.* **45:** 302- 309.

Joshi SV, Singh HP, Nair KNG and Mathew TV (1975). Regional variations in the farinographic and other quality characteristics of Kalyan Sona variety of wheat. *J. Food Sci. Technol.* **12:** 231- 233.

Juneja PK, Kawatra BL and Bajaj S (1980). Nutritive value of triticale and the effects of its supplementation to wheat and Bengal gram (*Cicer aeritinum*) flour. *J. Food Sci.* **45:** 328- 330 & 340.

Kadam ML, Salve RV, Mehrajfatema ZM and More SG (2012).Development and evaluation of composite flour for Missi roti / chapatti. *J. Food Proc. Technol.* **3:** 134-140.

Kahlon TS (2009). Rice bran: production, composition, functionality and food applications, physiological benefits. In: *Fiber Ingredients: food applications and health benefits*, Susan Sungsoo Cho and Priscilla Samuel (Eds.), pp 305- 321

Kameshwara Rao G, Malathi MA and Vijayaraghavan PK (1964). Preservation and packaging of Indian foods. I. Preservation of chapaties. *Food Technol.* **18:** 108- 110.

Kameshwara Rao G, Malathi MA and Mohan MS (1969). Preservation and packaging of Indian foods. III. *Parottas. J. Food Sci.Technol.* **6:** 279- 281.

Kannur SB, Eapen KC, Rao GK, Vijayraghavan PK and Nath H (1972). Heat processing of ready-to-eat foods in flexible pouches. I. stuffed *parottas* and cutlet. *J. Food Sci. Technol.* **9:** 179- 184.

Kannur SB, Prahlad K, Subramanian Y and Sharma TR (1985).A mechanical device for flattening of dough for making chapaties. *J. Food Sci.Technol.***22:** 60-61.

Kannur SB, Premavalli KS, Arya SS, Parihar DB and Nath H (1974).Studies on flavouring compounds of chapaties. I. Carbonyls. *J. Food Sci.Technol.* **11:** 5- 8.

Kannur SB, Ramanuja MN and Parihar DB (1973). Studies on the browning of heat processed chapaties and allied products in relation to their chemical constituents. *J. Food Sci. Technol.* **10:** 64- 67.

Kapoor S, Ranote PS and Sharma S (2015). Bioactive components and quality assessment of *jamun* (*Syzygium cumini* L.) powder supplemented chapatti. *Indian J. Sci. Technol.* **8:** 287- 295.

Karim AA, Norziah MH and Seow CC (2000). Methods for the study of starch retrogradation. *Food Chem.* **71:** 9- 36.

Katyal M, Virdi AS, Kaur A, Singh N, Kaur S, Ahlawat AS and Singh AM (2016). Diversity in quality traits amongst Indian wheat varieties. I: Flour and protein characteristics. *Food Chem.* **194:** 337- 344.

Kaur A (2009). Development of the ready mix formulation for "missi roti" a traditional flat bread. Poster presentation at *AACC International Annual Meeting*, 2009. American Association for Cereal Chemists, Baltimore, Maryland, USA.***https://www.scribd.com/.../*****Cereal-Foods-World-Supplement-54-*****4***

Kaur A, Bajaj M and Sidhu JS (1990), Studies on debranning of Punjab wheats I. Effect on chemical composition, *J. Plant Sci. Res.* **6:** 37-39.

Kaur A, Singh N, Shlawat AK, Kaur S, Singh AM, Chauhan H and Singh GP (2013). Diversity in grain, flour, dough and gluten properties amongst Indian wheat cultivars varying in high molecular weight subunits (HMW-GS). *Food Res. Int'l.* **53:** 63- 72.

Kaur H and Bajwa U (2000). Effect of heat refreshening on quality of stored chapattis containing buttermilk. *J. Food Sci. Technol.* **37:** 211- 216.

Kaur M, Mann SK, Hira CK and Bajaj S (1988). Effect of musk melon (*Cucumis melo*) seed supplementation on the nutritive value of wheat chapati. *J. Food Sci. Technol.* **25:** 263- 266.

Kaur M and Mehta U (1992).Nutritional and organoleptic characteristics of chapatis and paranthas supplemented with ricebean (*Vigna umbecullata*). *Bull. Grain Technol.* **30:** 246- 253.

Kaur RD and Hira CK (1988). Supplementary effect of *aestivum* wheat (*Triticum aestivum* L.), Bengal gram (*Cicer arietinum* L) and soya bean (*Glycine max* L) on protein quality of durum wheat (*Triticum durum* Desf). *J. Sci. Food Agric.***46:** 201- 209.

Kay DE (1979). *Food legumes.* Tropical Products Institute, London.

Kenny S, Wehrle K, Dennehy T and Arendt EK (1999). Correlations between empirical and fundamental rheology measurements and baking performance of frozen bread dough. *Cereal Chem.* **76:** 421- 425.

Khaliduzzaman M, Shams-Ud-Din M and Islam MN (2010). Studies on the preparation of chapatti and biscuit supplemented with potato flour. *J. Bangladesh Agril. Univ.* **8:** 153- 160.

Khalil AH, Mansour EH and Dawoud, FM (2000). Influence of malt on rheological and baking properties of wheat-cassava composite flours. *LWT- Food Sci. Technol.* **33:** 159-164.

Khalil JK and Chughtai MID (1982). Composition and Nutritive value of wheat bread supplemented with peanut flour. *Pakistan J. Scientific Res.* **34:** 149- 156.

Khan AN, Ahmed A, Bhatti MS, Randhawa MA, Ahmad A and Yousaf AA (2011). Effect of additives on the shelf-life extension of chapatti. *Food Sci. Technol. Res.* **17:** 203- 208.

Khan MA and Eggum BO (1978). Effect of baking on the nutritive value of Pakistani bread. *J. Sci. Food Agric.***29:** 1069- 1075.

Khan MA, Mahesh C, Semwal AD and Sharma GK (2015). Effect of spinach powder on physicochemical, rheological, nutritional and sensory characteristics of chapati premixes. *J. Food Sci. Technol.* **52:** 2359- 2365.

Khan MA, Mahesh C, Semwal AD, Sharma GK, Srihari SP, Jayaprahash C and Srihari KA (2014). Studies on shelf-life enhancement of potato stuffed parotha using thermal processing. *J. Food Sci. Technol.* **51:** 2190- 2196.

Khan MA, Semwal AD and Sharma GK, Khalid M, Srihari KA et al (2006). Development of chapatti/parotha with long shelf-life and the process for preparing the same. *Indian Patent* (Filed). No.2124/DEL/2006.

Khan MA, Semwal AD, Sharma GK, Mahesh C and Harilal PT (2013). Development and storage stability of spinach chapaties. *Int'l. J. Advanced Food Sci. Technol.* **1:** 12- 19.

Khan MA, Semwal AD, Sharma GK, Mahesh C, Nataraj S, Srihari KA and Bawa AS (2011). Development and evaluation of long shelf- life ambient stable chapaties without the use of chemical preservatives. *J Food Proc. Technol.* **2:** 1- 5.

Khan MI, Anjum FM and Akhtar S (2005a). Nutritional improvement of chapatti using soy hulls. *J. Res. (Science), BZ University, Multan, Pakistan* **16:** 57-63.**www.researchgate.net/publication**.

Khan MI, Anjum FM, Hussain S and Tariq MT (2005b). Effect of soy flour supplementation on mineral and phytate contents of unleavened flat bread (chapatis). *Nutr. Food Sci.* **35:** 163-168.

Khan MI, Anjum MF, Zahoor T, Sarwar M and Wahab S (2009). Nutritional characterization of the wheat-soy unleavened flat bread by rat bioassay. *Sarhad J. Agric.* **25:** 73-80.

Khan MN, Robinson RJ, Wang M and Hoover WJ (1972).Effect of protein-rich flour and sodium steroyl-2-lactylate (SSL) on chapatti (Indian non-leavened bread). (Abstr.). *Cereal Sci. Today* **17:** 263.

Khan N, Zaman R and Elahi M (1986). Effect of processing on the phytic acid content of wheat products. *J. Agric. Food Chem.* **34:** 1010-1012.

Khetarpaul N and Goyal R (2008). Development, nutritional and sensory attributes of nutritious bread prepared by using combination of wheat, soy and rice. *Acta Alimentaria* **37:** 147-157.

Kloots W, Op den Kamp D and Abrahmse L (2004). *In vitro* iron availability from iron-fortified whole-grain wheat flour. *J. Agric. Food Chem.* **52:** 8132-8136.

Kotturi K and Kannampilly NJ (2013). Development and evaluation of shelf stable partially baked chapatti using thermal processing. *IRACST- Engineering Sci. Technol. Int'l. J.* **3:** 260- 267.

Koushika Maya G, Shwetha BG and Indrani D (2015). Effect of barley flour on rheological characteristics of dough, organoleptic, nutritional and storage characteristics of south Indian *parotta*. *Food Sci. Technol. Int'l.* **21:** 24-32.

Kulp K and Ponte JG (1981). Staling of white pan bread: fundamental causes. *Crit. Rev. Food Sci. Nutr.* **15:** 1- 48.

Kulshrestha VP (1985). History and Ethnobotany of wheat in India. *J. d'Agric. Trad. et de Bota. Appl.* **32**: 61-71.

Kumar M and Sushma (2015). Study of phosphorus profile in Indian wheat varieties. *Int'l. J. Sci. Res.* **4:** 918-922.

Kumar V, Sinha AK, Makkar HPS and Becker R (2010). Dietary roles of phytate and phytase in human nutrition: a review. *Food Chem.* **120:** 945-959.

Kundu M, Gulia N and Khatkar BS (2016). Diversity in quality traits of Indian wheat cultivars: grain, whole wheat flour and protein characteristics. *IJIRSET* **5:** 17590-17594.

Kundu M, Khatkar BS and Gulia N (2017). Assessment of chapatti quality of wheat varieties based on physicochemical, rheological and sensory traits. *Food Chem.* **226:** 95-101.

Kurhade A, Patil S, Sonawane SK, Waghmare JS and Arya SS (2016). Effect of banana peel powder on bioactive constituents and microstructural quality of chapatti: Unleavened Indian flat bread. *J. Food Measur.* **10:** 32-41.

Larmond E (1977). *Laboratory methods for sensory evaluation of foods*. Canada Department of Agriculture Publication, Ottawa.

Lathia DN and Koch M (1989). Comparative study of phytic acid content, *in-vitro* protein digestibility and amino acid composition of different types of flat breads. *J. Sci. Food Agric*. **47:** 353-364.

Lee S, Inglet GE and Carriere CJ (2004). Effect of Nutrim oat bran and flaxseed on rheological properties of cakes. *Cereal Chem*. **81:** 637- 642.

Leelavathi K, Rao PH and Shurpalekar SR (1986) Studies on the functional characteristics of differently milled whole wheat flour (*atta*). *J. Food Sci. Technol*. **23:** 10-14.

Lindell MJ and Walker CE (1984). Soy enrichment of chapaties made from wheat and non-wheat flours. *Cereal Chem*. **61:** 435- 438.

Liu RH (2007). Whole grain phytochemicals and health. *J. Cereal Sci*. **46:** 207- 219.

Luthra G and Sadana B (1995). Potential of improving protein quality of wheat chapatti by addition of *bathu* (*Chinopodium album*), carrot (*Daucus carota*) and radish (*Raphanus sativus*) leaves. *J. Food Sci. Technol*. **32:** 507- 509.

Lyallpur Report (1943-46). Progress report for the scheme for wheat milling and baking tests for the years ending 30th June, 1945 and 30th June, 1947. *A Rep. Agric. Coll. Res. Inst., Lyallpur.* **Cross reference**.

Macritchie F (1984). Baking quality of wheat flours. *Adv. Food Res*. **29:** 201- 277.

Madenci AB and Bilgicli N (2014). Effect of whey protein concentrate and butter milk powders on rheological properties of dough and bread quality. *J. Food Quality* **37:**117- 124.

Majeed A, Shah TH, Rehman S and Siddique MI (1987). Development of technique and formulation for *roti* production on automatic plants. *Pak. J. Agril. Sci*. **24:** 210- 216.

Manjula Sasikumar R (2015). Effect of supplementation with flaxseed powder on *roti* quality. *Int'l. J. Sci. Res*. **4:** 912- 914.

Manohar R and Rao G (2005). Process for improver premix for chapatti and related products.*USPatentNo20050260318A1*, CSIR,NewDelhi,India.

www.google.com/patents/US20050260318.

Manu BT and Prasada Rao UJS (2008). Influence of size distribution of proteins, thiol and disulfide content in whole wheat flour on rheological and chapatti texture of Indian wheat varieties. *Food Chem*. **110:** 88- 95.

Maqbool A, Aurangzeb RA, Habib NB and Khan I (1987). Nutritional and organoleptic evaluation of wheat *rotis* supplemented with soybean flour. *Pak. J. Sci. Ind. Res*. **30:** 615- 618.

Marathe SA, Machaiah JP, Rao BYK, Pednekar MD and Rao VS (2002). Extention of shelf-life of whole-wheat flour by gamma radiation. *Int'l. J. Food Sci. Technol*. **37:** 163- 168.

Marnett LF, Tenney RJ and Berry VD (1978). Methods for producing soy-fortified breads. *Cereal Sci. Today* **18:** 38- 43, 50.

Mason R and Hoseney RC (1980). Preliminary pearl millet roti method. *Annual report to the Agency for International Development*, Kansas State University, Manhattan. **Cross reference.**

Mathur VK, Siddiah, CH, Bhatia BS and Vijayaraghavan PK (1973). Canning of conventional Indian meals for defence requirement. *Indian Food Packer* **27:** 46- 51.

Matthews RH, Richardson G and Lichtenstein H (1969). Effect of lysine fortification on quality of chapattis and yeast bread. *Cereal Chem.* 46**:** 14- 15.

Matz SA (1962). *Food Texture*. Avi Pub. Co., Westport, Conn., USA.

McCance RA and Walsham CM (1948). The digestibility and absorption of the calories, proteins, purines, fat and calcium in wholemeal wheaten bread. *Br. J. Nutr.* **2:** 26- 41.

McIntosh JR (2008). *The Ancient Indus Valley: New perspective*, ABC-CLO, Inc., Oxford, England. Books.google.co.in >History > Ancient > General.

Mehdi V and Abrol YP (1972). Studies on chapatti quality: IV.Total and phytic phosphorus in whole meal, milling fractions and chapatties. *Bull. Grain Tech.* **10:** 187- 189.

Mehdi V, Joshi LM and Abrol YP (1973), Studies on chapatti quality. VI. Effect of wheat grains with Bunts on the quality of chapatties. *Bull. Grain Technol.* **11:** 195- 197.

Mehta A, Mandeep K, Gupta S K and Singh R P (2006). Physicochemical, rheological and baking characteristics of bread wheat as affected by farmyard manure and nitrogen management. *J. Res. PAU.* **43:** 263- 270.

Milner CK and Carpenter KJ (1969). Effect of wet heat-processing on the nutritive value of whole wheat protein. *Cereal Chem.* **46:** 426- 434.

Misra, BK (1998). Quality needs for Indian traditional products. In *Proceedings of the International Group Meeting on "Wheat Research Needs Beyond 2000 AD"* held at Directorate of Wheat Research, Karnal, India,12-14 August, 1997. Eds. Nagarajan S, Singh G and Tyagi BS. Narosa Publishing House, New Delhi, pp 313- 319.

Mohan D and Gupta RK (2011). Harness value-addition in bread wheat (*Triticum aestivum*) through genotype and location specificity in highly productive north-western Indo-Gangetic plains. *Indian J. Agric. Sci.* **81:** 433- 437.

Moran T and Pace J (1942). Digestibility of high-extraction wheatmeals. *Nature* **150:** 224- 226.

Morrison WR (1978). Wheat lipid composition. *Cereal Chem.* **55:** 548- 558.

Mulla MZ, Bhardwaj VR, Annapure US and Singhal RS (2010). Effect of damaged starch on acrylamide formation in whole wheat flour based Indian traditional staples, chapattis and *pooris*. *Food Chem.* **120:** 805- 809.

Muller HG (1973).*An Introduction to Food Rheology*, Heinemann, London.

Muller HG (1970). Colloidal aspects of flour testing. *Chem. Ind.*p 1248.

Munro IC, Morrison AB and Myer M (1969). Fish protein concentrate as a supplement to cereal diets. *J. Am. Diet. Assoc.* **54:** 398- 400.

Murty GS and Austin A (1963). Studies on the quality characters of Indian wheats with reference to the mixability of their flours with the flours of other food grains and tubers for making chapaties. *Food Sci.* **12:** 64- 68.

Murthy NK, Annapurani S, Premjyothi P, Rajah J and Shubba K (1985). Bioavailability of iron by *in-vitro* method- 1- from selected foods and effect of fortification, promoters and inhibitors. *Indian J. Nutr. Diet.* **22:**68- 71.

Mustafa AI, Al-Wesseli MS, Al-Basha OM and Al-Amir RH (1986). Utilization of cowpea flour and protein isolate in bakery products. *Cereal Foods World* **31:** 756-759.

Nadeem M, Anjum FM, Khan MR, Shabbir MA, Saeed M and Ahmed A (2015). Immunological response of wheat varieties in relation to chapatti-making characteristics. *Food and Agricultural Immunology*, 2015:

http: // dx.doi.org/10.1080/09540105.2014.998632

Nagarajaiah SB and Prakash J (2011). Chemical composition and antioxidant potential of peels from three varieties of banana. *Asian J. Food Agro-Industry.* **4:** 31- 46.

Nandini CD and Salimath PV (2001). Carbohydrate composition of wheat, wheat bran, sorghum and bajra with good *chapatti/roti* (Indian flat bread) making quality. *Food Chem.* **73:** 197- 203.

Nanjappa C, Jagannath JH and Arya SS (1999). Thermal analysis of retrogradation of *phulka* and *chapatti*. *Indian Food Packer* **53:** 16- 20.

Narwal S, Kumar D, Sheoran S, Verma RPS and Gupta RK (2017). Hulless barley as a promising source to improve the nutritional quality of wheat products. *J. Food Sci. Technol.* **54:** 3638- 2664.

Nath N, Singh S and Nath HP (1957). Studies on the changes in the water soluble carbohydrates of chapaties during aging. *Food Res.* **22:**25- 31.

Navnidhi, Panghal A and Khatkar BS (2007). Process optimization for test baking of chapatti. *Indian Food Packer* **61:** 98 – 102.

Navnidhi, Panghal A, Singh A and Khatkar BS (2009). Quality assessment of Indian wheat varieties for steam leavened flat bread. *Annals Agri-Bio Res.* **15:** 195-200.

Nielsen MM and Hansen A (2008). Stability of vitamin E in wheat flour and whole wheat flour during storage. *Cereal Chem.* **85:** 716- 720.

Ocloo FCK, Bansa D, Boatin R, Adom T and Agbemavor WS (2010). Physicochamical, functional and pasting characteristics of flour produced from jackfruits (*Artocarpus heterophyllus*) seeds. *Agric. Biol. J. N. Am.* **1:** 903- 908.

Okaka JC and Potter NN (1979). Physicochemical and functional properties of cowpea powders processed to reduce beany flavour. *J. Food Sci.* **44:** 1235- 1240.

Ozdal T, Capanoglu E and Altay F (2013). A review on protein-phenolic interactions and associated changes. *Food Res. Int'l.* **51:** 954- 970.

Padwal Desai SR, Ghanekar AS, Thomas P and Sreenivasan A (1973). Heat radiation combination for control of mold infection in harvested fruits and processed cereal foods. *Acta Aliment.Acad.Sci.Hung.* **2:** 189- 207. **Cross reference.**

Pande S, Sakhare SD, Bhosale MG, Haware DJ and Inamdar AA (2017). *Atta* (whole wheat flour) with multi-whole-grains: flour characterization, nutritional profiling and evaluation of chapatti making quality. *J. Food Sci. Technol.* **54:** 3451- 3458.

Panghal A, Navnidhi C and Khatkar BS (2019). Characterization of Indian wheat varieties for chapatti (flat bread) quality. *J. Saudi Soc. Agric. Sci.* **18:** 107- 111.

Parihar DB and Chatterji AK (1956). X-ray diffraction studies of chapatti during cooking and storage. *J. Sci. Ind. Res.* (New Delhi). **15 C:** 115- 117.

Parimala KR and Sudha ML (2012). Effect of hydrocolloids on the rheological, microscopic, mass transfer characteristics during frying and quality characteristics of *puri*. *Food Hydrocoll.* **27:** 191-200.

Parimala KR and Sudha ML (2015). Wheat-based traditional flat breads of India. *Cri. Rev. Food Sci. Nutr.* **55:** 67- 81.

Pathania S, Kaur A and Sachdev PA (2017). Chickpea flour supplemented high protein composite formulation for flatbreads: Effect of packaging materials and storage temperature on the ready mix. *Food Packaging and Shelf life* **11:** 125- 132.https://doi.org/10.1016/j.fpsl.2017.01.006.

Payne PL, Nightingale MA, Krattiger AF and Holt LM (1987). The relationship between HMW glutenin composition and the bread making quality of British grown wheat varieties. *J. Sci. Food Agric.* **40:** 51- 65.

Pelshenke PA (1933). A short method for the determination of gluten quality of wheat. *Cereal Chem.* **10:** 90. **Cross reference.**

Pirayeshfar B, Rao PH and Sai Manohar R (2001). Effect of surfactants on quality of chapatti. *Seed and Plant* **17:** 21- 31.

Plaami SP (1997). Content of dietary fibre infoods and its physiological effects. *Food Reviews Int'l.* **13:** 29- 76.

Pomeranz Y (1988). *Wheat: Chemistry and Technology*, American Association of Cereal Chemists, St. Paul, MN.

Popli S and Dhindsa KS (1980). Quality characteristics of improved wheat varieties of Haryana and Punjab. *Bull. Grain Technol.* **18:** 10 – 16.

Prabhasankar P (2002). Electrophoretic and immunochemical characteristics of wheat protein fractions and their relationship to chapatti-making quality. *Food Chem.* **78** (1): 81- 87.

Prabhasankar P, Indrani D, Jyotsna R and Venkateswara Rao G (2003). Scanning electron microscopic and electrophoretic studies of the baking process of South Indian *parotta* – an unleavened flat bread. *Food Chem.* **82:** 603- 609.

Prabhasankar P, Indrani D, Jyotsna R and Venkateswara Rao G (2004). Influence of enzymes on rheological, microstructure and quality characteristics of *parotta*- An unleavened flat bread. *J. Sci. Food Agric.* **84:** 2128- 2134.

Prabhasankar P, Manohar R and Gowda LR (2002). Physicochemical and biochemical characteristics of selected wheat cultivars and their correlation to chapatti making quality. *Eur. Food Res. Technol.* **214:** 131- 137.

Prabhasankar P and Rao PH (2001). Effect of different milling methods on chemical composition of whole wheat flour. *Eur. Food Res Technol.* **213:** 464- 469.

Prabhasankar P and Sai Manohar R (2002). Development of enzyme-linked immunosorbent assay for evaluation of chapatti-making quality of wheat varieties. *J. Agric. Food Chem.* **50:** 7455- 7460.

Prabhavathi C, Rao PH and Shurpalekar SR (1976). Bread and chapatti making quality of Indian durum wheats. *J. Food Sci. Technol.* **13:** 313 – 317.

Prabhavathi T and Narasinga Rao BS (1979). Effects of domestic preparation of cereals and legumes on ionisable iron. *J. Sci. Food Agric.* **30:** 597- 602.

Prasad NN, Santhanum K and Rama Rao MV (1987). Protein quality of chapathies preserved with sorbic acid and stored for six months. *J. Sci. Food Agric.* **39:** 73- 77.

Preethi B M and Chimmad BV (2014). Development of ready to use omega-3 enriched chapati flour through value addition to linseed (*Linum usitatissimum*). *Int'l. J. Food Agric. Vet. Sci.* **4:** 160- 164.

Prinyawiwatkul W, Mcwatters KH, Beuchat LR, Phillips RD and Uebersak MA (1996). Cowpea flour: a potential ingredient in food products. *Crit. Rev. Food Sci, Nutr.* **36:** 413 - 436.

Proceedings of the 55th All India Wheat and Barley Workersmeet held at CCSHAU, Hisar from 21-24 August, 2016.All India Coordinated Wheat and Barley Improvement Project.

Punder de K and Pruimboom L (2013). The dietary intake of wheat and other cereal grains and their role in inflammation. *Nutrients* **5:** 771- 787. **doi:10.3390/ nu5030771.**

Punia D, Yadav SK, Gupta M and Khetarpaul N (2004). Nutrient composition of amaranth (*Amaranthus tricolor*) and *kondhara(Digeria arvensis*) leaves and their products. *J. Food Sci. Technol.* **41:** 563- 566.

Purhagen JK, Sjoo ME and Eliasson A-C (2012). Fibre-rich additives- the effect on staling and their function in free-standing and pan-baked bread. *J. Sci. Food Agric.* **92:** 1201- 1213.

Qarooni J, Ponte JG and Posner ES (1992). Flat breads of the world. *Cereal Foods World* **37:** 863- 865.

Raghavan CV, Sai MR and Haridas Rao P (1994). Effect of key factors on mechanical sheeting characteristics of chapatti dough made from whole wheat flour and resultant *atta. J. Food Sci. Technol.* **31:** 202- 206.

Rahim A, Vatsala CN and Shurpalekar SR (1993). Development of a laboratory method for preparation of *Nan. J. Food Sci. Technol.* **30:** 114- 117.

Raja KCM (1986). Texture characteristics of cassava-wheat dough and sensory properties of chapathi. *J. Food Sci. Technol.* **23:** 77- 79.

Ram, BP and Nigam SN (1979). The quality of wheat flour: intervarietal quantitative differences between gluten components. *J. Food Sci. Technol.* **16:** 189- 192.

Ram, BP and Nigam SN (1981). Puffing and textural characteristics ofchapatti in relation to varietal differences in gluten composition. *J. Food Sci.* 47**:** 231- 233.

Ram S (2003). High molecular weight glutenin subunit composition of Indian wheats and their relationships with dough strength. *J. Plant Biochem. Biotechnol.* **12:** 151- 155.

Ram S and Mishra B (2007).Advances in improving wheat quality for different end-products in India. *J. Wheat Res.* **1:** 86- 96.

Ram S and Mishra B (2010). Cereals – Processing and nutritional quality. New India Publishing Agency, New Delhi, India.

Ramaswamy H and Marcotte M (2006). Food processing- principles and applications. Publisher: CRC Press, Boca Raton, Fl., USA.

Randhawa, M. S. (1980). *A history of Agriculture in India*, **Vol. 1**, Indian Council of Agricultural Research, New Delhi.

Ranga Rao GCP, Rao PH, Leelavathi K and Shurpalekar SR. (1986). Effect of heat developed during grinding wheat in a disc mill on some chemical, rheological and chapati making characteristics of flour. *J. Food Sci. Technol.* **23:** 29 – 32.

Rao KG, Malathi MA and Vijayaraghavan PK (1964). Preservation and packaging of Indian foods.I. Preservation of chapaties. *Food Technol.* **18:** 108- 112.

Rao KG, Malathi MA and Vijayaraghavan PK (1966). Preservation and packaging of Indian foods.II. Storage studies on preserved chapaties. *Food Technol.* **20:** 94- 99.

Rao PH (1982). Studies on chapatti and similar traditional foods. *Ph.D. Thesis*, University of Mysore, Mysore, India.

Rao PH (1993). Chapatis and related products. In *Encyclopedia of food science, food technology and nutrition*, Ed. Hui Y. H., **Vol. 2**, pp 795-801, Academic Press, London.

Rao PH, Leelavathi K and Shurpalekar SR. (1983). Comparative studies on Atta (Whole Wheat Flour) and resultant Atta a by-product of roller flour milling industry. *J. Food Sci. Technol.* **20:** 5- 8.

Rao PH, Leelavathi K and Shurpalekar SR (1986a). Objective measurements of the consistency of chapatti dough using a research water absorption meter. *J. Texture Studies* **17:** 401- 420.

Rao PH, Leelavathi, K and Shurpalekar, SR(1986b). Test baking of chapatti – Development of a method. *Cereal Chem.* **63:** 293 – 303.

Rao PH, Leelavathi, K and Shurpalekar, SR (1989). Effect of damaged starch on the chapatti-making quality of whole wheat flour. *Cereal Chem.* **66:** 329- 333.

Rao TSS, Kuttikrishnan C, Rama Rao MV and Nath H (1974). Effect of feeding chapaties preserved with sorbic acid on growth and reproduction of Albino rats. *Def. Res. J.* **24:** 57- 60.

Rashid J and Hawthorn J (1974). Triticale as a cereal for chupatty baking. (Abstract). *Cereal Sci. Today*. **19:** 414.

Rasool G, Anjum FM and Butt MS (2005). Bioevaluation of oilseed enriched wheat chapatties. *Int'l. J. Agri. Biol.* **7:** 663- 667. **http://www.ijab.org.**

Rasper VF and Walker CE (2000). Quality evaluation ofcereal and cereal products. In:*Handbook of Cereal Science and Technology, Second Edition*, Kulp K and Ponte JG (Eds), Marcel Dekker, Inc., New York, p 505- 537.

Rathod KL and Williams SW (1970). Consumer acceptance of chapattis containing soybean flour. *Indian J. Agric. Econ.* **25:** 68- 72.

Rathod KL and Williams SW (1973). Acceptance of soybean-enriched wheat flour for use in chapattis. *Indian J. Nutr. Diet.* **10:** 18- 25.

Rawat A, Singh G, Mital BK and Mittal SK (1994). Effect of soy-fortification on quality characteristics of chapatis. *J. Food Sci. Technol.* **31:** 114- 116.

Ray A and Dasappa I (2017). Effect of natural preservatives on the rheological, physico-sensory, microbiological characteristics and shelf life of South Indian parotta. *J. Food Measure*. **11:** 660- 666.

Rehman S U, Ahmad MM, Bhatty IA, Shafique R, Din GMU and Murtaza MA (2006a). Effect of pearling on physicochemical, rheological characteristics, and phytate content of wheat-sorghum flour. *Pak. J. Bot.* **35:** 711- 719.

Rehman SU, Paterson A, Hussain S, Bhatti IA andShahid MAR (2006b). Influence of detoxified Indian vetch (*Lathyrus sativus* L.) on sensory and protein quality characteristics of composite flour chapatti. *J. Sci. Food Agric.* **86:** 1172- 1180.

Rehman SU, Paterson A and Piggott JR (1997).Quantification of total flour protein and its relationship to rheological characteristics related to chapatti. *Sci. Int'l.* **9:** 55- 59.

Rehman SU, Paterson A and Piggott JR (2006c). Optimization of chapatti quality using British wheat cultivar flours. *Int'l. J. Food Sci. Technol.* **41:** 1- 7.

Rehman SU, Paterson A and Piggott JR (2007). Optimization of flours for chapatti preparation using a mixture design. *J. Sci. Food Agric.* **67:** 425- 430.

Reinhold JG, Faradji B, Abadi P and Ismail-Beigi F (1981). An extended study of the effect of Iranian village and urban flat breads on the mineral balances of two men before and after supplementation with vitamin D. *Ecol. Food Nutr.* **10:** 169-177. **Cross reference.**

Revanappa SB, Bhagwat SG and Salimath PV (2007). Studies on pentosans in Indian wheat (*Tritcum aestivum*) varieties in relation to chapati making quality. *Food Chem.* **104:** 896- 902.

Rice-Evans CA, Miller NJ and Paganga G (1997). Antioxidant properties of phenolic compounds. *Trend Plant Sci.* **2:** 152- 159.

Robert SD, Ismail AA and Rosli WIW (2015). Reduction of postprandial blood glucose in healthy subjects by buns and flat breads incorporated with fenugreek seed powder. *Eur. J. Nutr.* **55:** 2275- 2280. **DOI: 10.1007/s 00394-015-1937-4.**

Rogers DE, Zeleznak KJ, Lai CS and Hoseney RC (1988). Effect of native lipids, shortening and bread moisture on bread firming. *Cereal Chem.* **65:** 398- 401.

Rosell CM and Gomez M (2007). Frozen dough and partially baked bread: an update. *Food Rev Int'l.* **23:** 303- 319.

Roy Choudhuri RN, Joseph AA, Joseph K, Narayana Rao M, Swaminathan M, Sreenivasan A and Subrahmanyan V. (1963). Preparation and chemical composition of potato flour from some varieties of potato. *Food Sci.* **12:** 251- 253.

Rubenthaler GL and Faridi HA (1982).Laboratory dough molder for flat breads. *Cereal Chem.* **59:** 72- 73.

Sabir A, Chaudhry MS and Sattar A(1982). Effect of simultaneous removal of fines and bran on certain minerals, phytic acid and chapati quality of wheat flour. *Lebens. Wiss. Technol.* **15:** 255- 257. **FST Abstracts.**

Sadana B, Hira C and Kang H (1989).Sensory evaluation of chapattis prepared from different levels of infested wheat flour. *Beverage & Food World,* **April/June:** 19-20.

Sadasivan G, Rai SC and Austin A (1971). *J. Food Sci. Technol.* **8:** 7.

Sadasivan G, Sundaram SS and Austin A (1976). Thurstone-Moisteller model for palatability of chapattis of some improved bread wheats. *J. Food Sci. Technol.* **13:** 333- 334.

Sakhare SD, Inamdar AA, Soumaya C, Indrani D and Venkateswara Rao G (2014). Effect of flour particle size on microstructural, rheological and physico-sensory characteristics of bread and South Indian *parotta*. *J. Food Sci. Technol.* **51:** 4108- 4113.

Sarojini G, Nirmala G and Chaturvedi A (1996 a). Studies on utilization of Rajmah (*Phaseolus vulgaris*) flour in some Indian traditional products. *J. Food Sci. Technol.* **33:** 159- 161.

Sarojini G, Nirmalamma N and Chaturvedi A (1996 b). Product development and evaluation of whole grain and flour blends of Amaranthus paniculatus. *J. Food Sci. Technol.* **33:** 345- 347.

Savagaon KA, Dharkar SD and Sreenivasan A (1970). Radiation preservation of chapaties (Indian unleavened bread). *Food Technol.* **24:** 116- 118.

Sawhney B and KawatraBL(1986). Supplementary value of beet (*Beta vulgaris*), Knol-khol (*Brassica oleracea*) and turnip (*Brassica* rapa). *J. Food Sci. Technol.* **23:** 311- 315.

Saxena DC and Haridas Rao P (1995). Survey of the quality characteristics of *tandoori* dough and *tandoori roti*. *J. Food Sci. Technol.* **32:** 74- 76.

Saxena DC and Haridas Rao P (1996). Optimization of ingredients and process conditions for the preparation of *tandoori roti* using response surface methodology. Int'l. J. Food Sci. Technol. **31:** 345- 351.

Saxena DC and Haridas Rao P (1996). Effect of different baking modes on the physicochemical and sensory characteristics of *tandoori roti*. *Z Lebensm Unters Forch.* **203:** 262- 267.**FSTA Abstracts.**

Saxena DC, Haridas Rao P and RaghavaRao KSMS (1995). Analysis of modes of heat transfer in *tandoor* oven. *J. Food Eng.* **2:** 209- 217.

Saxena DC, Prasada Rao UJS and Haridas Rao P (1997). Indian wheat cultivars: correlation between quality of gluten proteins, rheological characteristics of dough and *tandoori roti* quality. *J. Sci. Food Agric.* **74:** 265- 272.

Saxena DC, Salimath PV and Haridas Rao P (2000). Indian wheat cultivars: their carbohydrate profile and its relation to *tandoori roti* quality. *Food Chem.* **68:** 185- 190.

Saxena DC and Rao HP (1997). Influence of grinding methods on some physicochemical characteristics of whole wheat flour and its suitability for the preparation of *tandoori roti*. *Sciences- des-Aliments* **17:** 183- 191. **FSTA Abstracts**.

Saxena DC and Rao HP (2000). Effect of damaged starch on the pasting characteristics and *tandoori roti*-making quality of whole wheat flour. *Sciences- des-Aliments* **20:** 591- 602. **FSTA Abstracts.**

Schneeman BO (1978). Effect of plant fibre on lipase, trypsin and chymotrypsin activity. *J Food Sci.* **43:** 634- 635.

Schofield RK and Scott Blair GW (1932).The relationship between viscosity, elasticity and plastic strength of soft material as illustrated by some mechanical properties of flour doughs. I. *Proc.Roy. Soc. (London).* **A 138:** 707-719.**Cross reference.**

Seibel W (2011). Future of Flours: Composite flours.

www:muehlenchemie.de/downloads-future-of-flour/FOF_Kap_16 pdf.

Sekhon KS, Saxena AK, Randhawa SK and Gill KS (1980). Use of triticale for bread, cookie and chapatti making. *J. Food Sci. Technol.* **17:** 233- 235.

Seleem HA and Omran AA (2014). Evaluation of quality of one layer flat bread supplemented with beans and sorghum baked on hot metal plate. *Food Nutr. Sciences.* **5:** 2246- 2256. d**oi: 10.4236/fns.2014.522238**.

Shahidi HM and Muller HG (1975). Fate of thiamine during baking of Pakistani and traditional Iranian bread. *J. Sci. Food Agric.* **26:** 542- 543.

Shahzadi N, Butt MS, Rehman SU and Sharif K (2005). Rheological and baking performance of composite flours. *Int. J. Agri. Biol.* **7:** 100- 104.

Shaikh IM, Ghodke SK and Ananthanarayan L (2007). Staling of chapatti (Indian unleavened flat bread). *Food Chem.* **101:**113-119.

Shaikh IM, Ghodke SK and Ananthanarayan L (2008). Inhibition of staling in chapatti (Indian unleavened flat bread). *J. Food Process. Preserv.* **32:** 378-403.

Shanthala M and Prakash J (2005). Acceptability of curry leaf (*Murraya koinigii*) incorporated products and attitude towards consumption. *J. Food Process. Preserv.* **29:** 33- 44.

Sharavathy MK, Urooj A and Puttaraj S (2001). Nutritionally important starch fractions in cereal based Indian food preparations. *Food Chem.* **75:** 241- 247.

Sharma A, Rao VS, Bhagwat SG and Bapat MM (2004). Elongation curve for evaluating Chapati making quality of whole wheat flour and its relation with other quality parameters. *J. Food Sci. Technol.* **41:** 160-164.

Sharma B, Gujral HS and Solah V (2017). Effect of incorporating finger millet in wheat flour on mixolab behavior, chapatti quality and starch digestibility. *Food Chem.* **231:** 156- 164.

Sharma H and Bains GS (1976). Effect of grinding on the quality of whole wheat meal. *J. Food Sci. Technol.* **13:** 92-93.

Sharma HR and Chauhan GS (2000). Physicochemical and rheological quality characteristics of fenugreek (*Trigonella foenum-graecum* L.) supplemented wheat flour. *J. Food Sci. Technol.* **37:** 87- 90.

Sharma M, Sandhir R, Singh A, Kumar P, Mishra A, Jachak S, Singh SP, Singh J and Roy J (2016). Comparative analysis of phenolic compounds: characterization and their biosynthesis genes between two diverse bread wheat (*Triticum aestivum*) varieties differing for chapatti (unleavened flat bread) quality. *Front. Plant Sci.* **7:**1- 17. **Doi.10.3389/fpls.2016.01870.**

Sharma P and Gujral HS (2014a). Antioxidant potential of wheat flour chapattis as affected by incorporating barley flour. *LWT- Food Sci. Technol.* **56:** 118- 123.

Sharma P and Gujral HS (2014b). Antistaling effects of β-glucan and barley flour in wheat flour chapatti. *Food Chem.* **145:** 102- 108.

Sharma R, Rawat A, Misra BK and Nagrajan S (2012). Distribution of grain hardness in Indian wheat varieties and landraces. *Wheat Information Service* **14:** 1- 8 (**www.shigen.nig.ac.jp/ewis**).

Sharma RD (1984). Hypocholestrolemic activity of fenugreek (*T. foenum graecum*). An experimental study in rats. *Nutr. Rep. Int'l.* **30:** 222- 231.

Sharma S, Bajwa U and Nagi HPS (1999). Rheological and baking properties of cowpea and wheat flour blends. *J. Sci. Food Agric.* **79:** 657- 662.

Sharma S, Nagi HPS and Sekhon KS (1988). Effect of blending laboratory sprouted grains on milling and baking properties of wheat. *J. Food Sci. Technol.* **25:** 330- 334.

Sharma S, Sekhon KS and Nagi HPS (1995). Legume supplemented flat bread: Nutritive value, textural and organoleptic changes during storage. *J. Food Process .Preserv.* **19:** 207- 222.

Shewry PR (2009). Wheat. *J. Exp. Bot.* **60:** 1537- 1553.

Shewry PR and Hey SJ (2015). The contribution of wheat to human diet and health. *Food and Energy Security* **4:** 178- 202.

Shenoy HA and Prakash J (2002). Wheat bran (*Triticum aestivum*): composition, functionality and incorporation in unleavened bread. *J. Food Quality* **25:** 197- 211.

Shurpalekar SR, Kumar GV, Rao GV, Ranga Rao GCP, Rahim A and Vatsala CN (1976). Physico-chemical, rheological and milling characteristics and bread and chapati making quality of Indian wheats. *J. Food Sci. Technol.* **13:** 79 – 83.

Shurpalekar SR and Prabhavathi, C (1976). Brabnder Farinograph, Research Extensometer and Hilliff Chapati Press as tools for standardization and evaluation of chapatti dough. *Cereal Chem.* **53:** 457 – 469.

Shurpalekar SR and Shukla VK (1992). Characteristics of *puri* dough and *puri* based on wheat and composite flours. *J. Food Sci. Technol.* **29:** 153- 158.

Shyamala G and Kennedy BM (1962). Protein value of chapatis and *puris. J. Am. Diet. Assoc.* **41:** 115- 117.

Siddique MI, Awan JA, Rehman SU, Gilani AH and Ali M (1996). Evaluation of composite flour based on wheat (*Triticum aestivum*) and chickpea (*Cicer arietinum*) for chapatti making. *Science Int'l.* **8:**165- 168.

Sidhu JS, Caceres PG and Behbehani M (1997). Measurement of starch properties during staling of Arabic flat bread. *Starch* **49:** 180- 186.

Sidhu JS, Harinder K, Kaur A and Ram MB (1990a). Functional and chapatti making properties of hull-less barley supplemented wheat flour. *J. Food Sci. Technol.* **27:** 313.

Sidhu JS, Seibel W and Bruemmer JM (1987). Semi-mechanized production of Indian unleavened flat bread (Chapati). *Gert. Mehl Brot* **41:** 334- 337.

Sidhu JS, Seibel W and Bruemmer JM (1989). Effect of shortening and surfactants on chapatti quality. *Cereal Foods World* **34:** 286- 290.

Sidhu JS, Seibel W and Meyer D (1990b). Gelatinization of starch during preparation of Indian unleavened flat breads. *Starch* **42:** 336- 341.

Sidhu JS, Seibel W, Brummer JM and Zwingelberg H (1988). Effect of flour milling conditions on the quality of Indian unleavened flat bread (chapatti). J. *Food Sci.* **53:** 1563-1565.

Silvestro RD, Loreto AD, Marotti I, Bosi S, Bregola V, Gianotti R, Quinn R and Dinelli G (2014). Effects of flour storage and heat generated during milling on starch, dietary fibre and polyphenols in stone ground flours from two durum-type wheats. *Int'l. J. Food Sci. Technol.* **49:** 2230- 2236. **DOI: 10.1111/ijfs.12536**

Singh M, Sharma NS and Khatkar BS (2006). End use quality of Indian wheats: Milling properties and composition. *J. Food Sci. Technol.* **43:** 322- 324.

Singh N, Harinder K, Sekhon KS and Kaur B (1991). Studies on the improvement of functional and baking properties of wheat-chickpea flour blends. *J. Food Process. Preserv.* **15:** 391- 402.

Singh N, Kaur SP, Kaur L and Sodhi NS (2005). Physicochemical, rheological and chapatti making properties of flours from some Indian potato cultivars. *J. Food Sci. Technol.* **42:**344- 348.

Singh P, Yadav N, Mishra PK and Sheikh S (2013). Utilization of rice bran for development of chapatti and its glycemic response in NIDDM patients. *Int'l. Res. J. Pharma. App. Sci.* **3:** 244- 248.

Singh R and Bailey CH (1940). A biochemical and technological study of Punjab wheat varieties. *Cereal Chem.* **17:** 160- 203.

Singh R and Sheoran IS (1972). Enzymic browning of whole wheat meal flour. *J.Sci. Food Agric.* **23:** 121- 125.

Singh RP (2001). Wheat quality requirements: A perspective. *India Grains* **3:** 17- 19.

Singh RP, Bansal G and Nanda GS (1999). Physicochemical and rheological characteristics of advanced wheat genotypes. *India Grains* **1:** 28- 34.

Singh RP, Sehgal KL, Saxena AK, Sekhon KS and Gill KS (1983). Physicochemical, rheological and baking characteristics of some promising strains of wheat. *J. Res, PAU.* **20:** 403 – 410.

Singhal RS and Kulkarni PR (1988). Review: Amaranthus- an underutilized resource. *Int'l. J. Food Sci. Technol.* **23:** 125- 139.

Sinha AC (1973). Quality of chapaties from Indian and imported wheats. *Bull Grain Technol.* **11:** 53- 59.

Sinha AC, Jain SK and Agrawal NS (1968). Studies on the physicochemical characteristics of improved varieties of Indian wheats. *Bull .Grain Technol.* **6:** 192- 201.

Sinha AC and Singh HP (1974). Studies on the farinographic and related characteristics of improved commercial varieties of Indian wheats. *Bull .Grain Technol.* **12:** 127- 131.

Smeds AL, Eklund PC, Sjoholm RE, Willfor SM, Nishibe S, Deyama T, Holmbom BR, et al (2007). Quantification of a broad spectrum of lignans in cereals, oilseeds and nuts. *J. Agric. Food Chem.* **55:** 1337- 1346.

Smitha S, Jyotsna R, Khyrunnisa B and Indrani D (2008). Effect of hydrocolloids on rheological, microstrutural and quality characteristics of *parotta*- unleavened flat bread. *J. Texture Studies* **39:** 267- 283.

Sood K, Dhaliwal YS, Kalia M and Sharma HR (1992). Utilization of hulless barley in chapatti making. *J. Food Sci. Technol.* **29:** 316- 317.

Sosulski FW, Elkowicz K and Reichert RD (1982). Oligosaccharides in eleven legumes and their air classified protein and starch fractions. *J. Food Sci.* **47:** 498- 502.

Sramkova Z, Gregova E and Sturdik E (2009). Chemical composition and nutritional quality of wheat grain. *Acta Chimica Slovaca* **2:** 115-138. **www.researchgate.net/publications/22835599.**

Sreeramulu G, Banerjee R, Bhardwaj A, Vaishnav PP. (2004). Understanding the functionality of wheat high molecular weight glutenin subunits (HMW.GS) in chapatti making quality. In *Proceedings of the 8th gluten workshop held on 8-10 September, 2003 in Viterbo, Italy*. Special Publication **No. 295**. Royal Society of Chemistry, Cambridge, UK. **Cross reference.**

Sridhar BS (1991). *Indian Pat.* No.177722. **Cross reference.**

Sridhar BS and Manohar B (2001). Optimization of the continuously extruded unleavened flat bread (*chapati*) process. *Eur. Food Res. Technol.* **212:** 477- 486.

Srivastava AK, Meyer D, Haridas Rao P and Seibel W (2002). Scanning electron microscopic study of dough and chapatti from gluten-reconstituted good and poor quality flour. *J Cereal Sci.* **35:** 119- 128.

Srivastava AK, Prasada Rao UJS, and Rao HP (2003). Studies on protein and its high-molecular-weight subunit composition in relation to chapatti making quality of Indian wheat cultivars. *J. Sci. Food Agric.* **83:** 225- 231.

Srivastava AK, Rao HP and Seibel W (1998). Role of gluten and its fractions in determining the chapatti making quality of whole wheat flour. Poster Abstracts; *4th International Food Convention Proceedings, 23-27 November, 1998, Mysore, India,* p 217.

Srivastava S, Hardainya D, Karoulia S and Chaudhary N (2015). Studies on utilization of buttermilk in chapatti making. *Int'l. J. Res. Emerg. Sci. Technol.* **2:** 30- 35.

Subrahmanyan V, Chandiramani SV, Rajagopalan R, Bhatia DS and Swaminathan M (1958). Utilization of tapioca flour and low fat groundnut flour admixed with wheat flour for the preparation of chapati and *poori. Food Sci*. **7:** 4- 6.

Sudha ML and Venkateshwara Rao G (2009). Influence of hydoxypropylmethyl cellulose on the rheological and microstructural characteristics of whole wheat flour dough and quality of *puri. J. Texture Studies* **40:** 172- 191.

Sudha ML, Vetrimani R and Leelavathi K (2007). Influence of fibre from different cereals on the rheological characteristics of wheat flour dough and on biscuit quality. *Food Chem.* **100:** 1365- 1370

Sultana A, Rehman MRT, Islam M, Rehman M and Alim MA (2014). Evaluation of quality of chapaties enriched with jackfruit seed flour and Bengal gram flour. *J. Environ. Sci. Technol. Food Technol.* **8:** 73- 78.

Supekar DT, Patil SR and Munjal SV(2005). Comparative study of some important *aestivum, durum* and *dicoccum* wheat cultivars for grain, flour quality and suitability for chapatti making characteristics. *J. Food Sci. Technol.* **42:** 488- 492.

Suribabu K, Lalitha Goverdhan T and Hemalatha KPJ (2014). Applications of partially purified α-amylase produced by Brevibacillus borstelensis R 1 in food industries. *World J. Pharma. Res.* **3:** 4989- 4995.

Survase A, Furtado A, Thengane R, Fox G, Taylor T and Henry R. (**13 May 2017 on line**). Comparison of chapatti and bread making quality of wheat genotypes. **http:// dx. Doi.org/10.1094/CCHEM-04-16-0090- R.**

Swaranjeet K, Maninder K and Bains GS (1982). Chapaties with leavening, and supplements: Changes in texture, residual sugars, and phytic phosphorus. *Cereal Chem.* **59:** 367- 372.

Syed HM, Rathi SD, More DR and Yasmin HZ (1993). Studies on improving the chapatti quality. *Indian Miller* **24:** 25- 30.

Syed HM, Rathi SD and Sawate AR (1991). Studies on improving chapatti quality. *Indian Miller* **22:** 26- 31.

Syed HM, Rathi SD and Sawate AR (1992). Hydrolysis of phytate in whole meal and *atta* chapaties. *Bull. Grain Technol.* **30:** 32- 37.

Taneja S, Gupta K, Wagle DS and Dhindsa KS (1983). Biological evaluation of wheat varieties. *J. Food Sci. Technol.* **20:** 319 – 320.

Tara KA, Haridas Rao P, and Bains GS (1969). Composition and rheological and baking quality of a by-product of wheat milling simulating whole wheat meal. *J. Sci. Food Agric.* **20:** 368- 373.

Tara KA, Usha MSM, and Bains GS (1972). Effect of lysine on dough and protein quality of whole wheat meal chapattis and leavened bread. *J. Agri. Food Chem.* **20:** 116- 118.

Tejaswi Kumar HC, Jeyarani T, Milind and Indrani D (2013). Effect of fats and oils on rheological, fatty acid profile and quality characteristics of South Indian parotta. *J. Am. Oil Chem. Soc.* **90:** 1523- 1531.

Teotia MS and Ramakrishna P (1984). Chemistry and technology of melon seeds. *J. Food Sci. Technol.* **21:** 332- 340.

The American Dietetic Association (1993). The health implications of dietary fibre. *J. Am. Diet. Assoc.* **93:** 1446- 1447.

Thompson LU (1993). Potential health benefits and problems associated with antinutrients in foods. *Food Res. Int'l.* **26:** 131- 149.

Thompson LU (1994). Antioxidant and hormone-mediated health benefits of whole grains. *Crit. Rev. Food Nutr.* **34:** 473- 497.

Thondre PS, Henry C J K (2009). High-molecular-weight barley β-glucan in chapatis (unleavened Indian flatbread) lowers glycemic index. Nutr. Res, **29:** 480- 486.

Tikoo S, Singh T, Abrol, YP and Sachar, RC (1973). Studies on polyphenol oxidase in wheat grain. *Cereal Chem.* **50:** 520- 528.

Tulyathan V, Tananuwong K, Songjinda P and Jaiboon N (2002). Some physicochemical properties of jackfruit (*Artocarpus heterophyllus* L.) seed flour and starch. *Science Asia* **28:** 37- 41.**Doi.10.2306/scienceasia1513-1874.2002.28.037.**

Uprety DC and Abrol YP (1972). Studies on chapatti quality: V. Palatability of chapaties.*Bull. Grain Technol.* **10:** 276- 282.

Uprety DC and Austin A (1969). Comparative study of the chemical composition and baking quality of 'Sonora 64' and '*sharbati* Sonora'. Indian J. Agric. Sci. **39:** 282- 285.

USDA (2016). National nutrient database for standard reference, release 28. *United States Department of Agriculture Nutrient Database (2016).* Flaxseed.**http://ndb.nal.usda.gov/.**

USDA (2016). National nutrient database for standard reference, release 28.*United States Department of Agriculture Nutrient Database (2016).* Flat bread naan. **http://ndb.nal.usda.gov/.**

USDA (2016). National nutrient database for standard reference, release 28, *United States Department of Agriculture Nutrient Database (2016)*, Wholegrain wheat flour, and chapatti. **http://ndb.nal.usda.gov/.**

USDA (2016). National nutrient database for standard reference, release 28, *United States Department of Agriculture, Nutrient Database (2016).* North Indian *Paratha.* **http://ndb.nal.usda.gov/.**

Vatsala CN, Saxena DC and Haridas Rao P (2001). Optimization of ingredients and process conditions for the preparation of *puri* using response surface methodology. Int'l. *J. Food Sci. Technol.* **36:** 407- 414.

Venkateshmurthy K, Jayaprakashan SG (1991). A device for pneumatic extrusion of dough into sheet or strands. No. NS-1/91-PAT dated 23-1-91. *Patents Unit, CSIR, New Delhi, India.* **Cross reference.**

Venkateshmurthy K, Jayaprakashan SG (1991). A dusting and cutting machine useful for chapatti and other similar products. No. NS-4/91-PAT dated 23-1-91. *Patents Unit, CSIR, New Delhi, India.* **Cross reference.**

Venkateshmurthy K, Jayaprakashan SG, Gadigeppa B, Indrajit M, Raghavarao KSMS, Sudha ML, Venkateswara RG (2008). An automatic device for pneumatic extrusion of dough into sheet and shaping, cutting of extruded sheet for preparation of *poory* and other similar Indian traditional foods. *Indian Patent application No: 135/Del/2010.* **Cross reference.**

Venkatesh Murthy K, Sai Manohar R and Rao HP (1998). Factors affecting sheeting characteristics of chapatti dough using pneumatic pressure extrusion. *J. Food Sci. Technol.* **35:** 122- 127.

Venkateshmurthy K, Sudha ML, Ravi R and Raghavarao KSMS (2015). Optimization of pneumatic sheet extrusion of whole wheat flour *poory* dough using response surface methodology. *J. Food Sci. Technol.* **52:** 4405- 4413.

Venkateswara Rao,G, Leelavathi K, Rao PH and Shurpalekar SR (1986). Changes in quality characteristics of chapatti during storage. *Cereal Chem.* **63:** 131- 135.

Venkateswara Rao G, Seibel W, Bretschneider F and Shurpalekar SR (1981). Preservation of chapattis by heat sterilization. *Getr. Mehl. Brot* **35:** 137-139. **Abstract.**

Verma NS, Mishra HN and Chauhan GS (1987). Preparation of full fat soy flour and its use in fortification of wheat flour. *J. Food Sci. Technol.* **24:** 259- 260.

Vidyasagar K and Arya SS (1981). Effect of sorbic acid on the chemical and rheological changes during mixing and chapati baking. *J Food Sci. Technol.* 1**8:** 62- 63.

Vijayakumar TP, Mohankumar JB and Jaganmohan R (2009). Quality evaluation of chapatti from millet flour blend incorporated composite flour. *Indian J. Nutr. Diet.* 46**:** 144- 155.

Vijayalakshmi D, Sharada D and Sastry M (1983). Protein quality of casein co-precipitate incorporated recipes. *Indian J. Dairy Sci.* **36:** 354-358.

Vijaya Rao D, Leela RK and Sankaran R (1979). Microbial studies on in-pack processed chapaties. *J. Food Sci. Technol.* **16:** 166- 168.

Waghray K, Gulla S and Bhanuteja K (2011). Sensory attributes of wheat (*Triticum* species) flour based "chapathis"incorporated with soy flour (*Glycine max*) and dates (*Pheonis dactylifera*). *J. Dairying Food & H.S.* **30:** 262- 266.

Walia V (2003). *Langar* gets a taste of technology. *The Tribune, Windows*, April 5, 2003, p 1.

Wani IA, Sogi DS, Sharma P and Gill BS (2016). Physicochemical and pasting properties of unleavened flat bread (chapatti) as affected by addition of pulse flour. *Cogent Food & Agriculture* 2: **DOI: 10.1080/23311932.2015.1124486.**

Watson SA (1997). Structure and composition. In: *Corn Chemistry and Technology*, SA Watson and PE Ramstad (Eds.). pp. 311- 349. Saint Paul, MN, USA: AACC. **Cross reference.**

Weiser H, Seilmeier W and Kieffer R (1994). Relationship between the amount of gluten protein types and the rheological properties of different wheat cultivars. In: *Gluten proteins 1993. Dermold, Germany: Assoc. Cereal Res.*, p 141-150.**Cross reference.**

Welch RM and Graham RD (1999). A new paradigm for world agriculture: meeting human needs - productive, sustainable, nutritious. *Field Crops Res.* **60:** 1- 10.

Welch RM and Graham RD (2004). Breeding for micronutrients in staple food crops from a human nutrition perspective. *J. Exp. Bot.* **55:** 353- 364.

Wijngaart van den A and Codling K (2013). Impact of flour fortification on common Asian wheat flour-based foods. *Cereal Foods World* **58:** 238- 245.

Wong S, Li C, Copeland L, Niu Q and Wong S (2015). Starch retrogradation: A comprehensive review. Compr. *Rev, Food Sci. Food Saf.* **14:** 568- 585.

Wu W, Huff HE and Hsieh F (2007).Processing and properties of extruded flaxseed-corn puff. *J. Food Process. Preserv.* **31:** 211-226.

Yadav DN, Patki PE, Khan MA, Sharma GK and Bawa AS (2008c). Effect of freeze-thaw cycles and additives on rheological and sensory properties of ready-to-bake frozen chapaties.*Int'l. J. Food Sci. Technol.* **43:** 1714- 1720.

Yadav DN, Patki PE, Mahesh C, Sharma GK and Bawa AS (2008b). Optimization of baking parameters of chapatti with respect to vitamin $\mathbf{B}_1$ and $\mathbf{B}_2$ retention and quality. *Int'l. J. Food Sci. Technol.* **43:** 1474-1483.

Yadav DN, Patki PE, Sharma GK and Bawa. AS (2008a). Effect of microwave heating of wheat grains on the browning of dough and quality of chapattis. *Int'l. J. Food Sci. Technol.* **43:** 1217- 1225.

Yadav DN, Patki PE, Sharma GK, Harilal PT and Bawa AS (2009a). Studies on the prevention of dough discolouration by chemical additives. *J. Food Sci. Technol.* **46:** 149- 152.

Yadav DN, Patki PE, Sharma GK, Srihari KA, Harilal PT and Bawa AS (2009b). Rheological and sensory quality of ready-to-bake chapatti during frozen storage. *J. Food Quality.***32:**436-451.

Yadav DN and Rajan A (2012). Fibres as additive for oil reduction in deep fat fried *poori. J. Food Sci. Technol.* **49:** 769- 773.

Yadav DN, Rajan A, Sharma GK and Bawa AS (2010). Effect of fiber incorporation on rheological and chapatti making quality of wheat flour. *J Food Sci. Technol.* **47:** 166- 173.

Yamazaki WT and Donelson JR (1983). Kernel hardness of some US wheats. *Cereal Chem.* **60**: 344-350.

Yasin M, Khan AH and Karimullah H (1965). Studies of the preservation of chapatis. *Pak. J. Sci. Ind. Res.* **8:** 126- 129.

Yasoda DP and Geervani P (1983). Effect of different heat treatments on losses of lysine and tryptophan in wheat products. *Indian. J. Nutr. Diet.* **20:** 356- 360.

Yasunaga T, Bushuk W and Irvine GN (1968). Gelatinization of starch during bread-baking. *Cereal Chem.* **45:**269- 279.

Zeleny I (1947). A simple sedimentation test for estimating the bread-making and gluten qualities of wheat flour. *Cereal Chem.* **24:** 465- 475.

Zeleznak KJ and Hoseney RC (1986). The role of water in retrogradation of wheat starch gels and bread crumb. *Cereal Chem.* **63:** 407- 411.